MUSHROOMS OF IDAHO
and the
PACIFIC NORTHWEST
Volume One

Best Regards !

E. Tylutki

10/96

MUSHROOMS OF IDAHO
and the
PACIFIC NORTHWEST

Volume One
Discomycetes

**Morels, False Morels, Fairy Cups, Saddle Fungi,
Earth Tongues, Truffles and Related Fungi
(Ascomycetes — Discomycetes)**

Edmund E. Tylutki

A NORTHWEST NATURALIST BOOK
University of Idaho Press
Moscow, Idaho
1993

The information in this book is accurate to the best of the author's knowledge. However, neither the author nor the publisher can accept responsibility for mistakes in identification or idiosyncratic reactions to mushrooms. It is certainly not necessary to eat mushrooms in order to enjoy them. People who choose to eat mushrooms do so at their own risk.

Published by the University of Idaho Press
Moscow, Idaho 83844-1107

ERRATA

Mushrooms of Idaho and the Pacific Northwest
Discomycetes

by E. E. Tylutki

Following are corrections to typographical errors and changes to scientific names currently in use.

61	to List of Species add: **Gyromitra melaleucoides** (Seaver) Pfister
62	to List of Species add: **Leptopodia epihippium** (Lev.) Boud.
65	key point 18b add: **Leptopodia epihippium**
	key point 19b **H. griseoalba** is: **H. griesoalba**
86	in List of Species:
	19 **Meladina lechitina** is: 19 **Miladina lechithina**
	20 **Melastiza chateri** is: 20 **Melastiza chatheri**
	31 **Pyronema omphaloides** is: 31 **Pyronema omphalodes**
	34 **Scutellinia umbrarum** is: 34 **Scutellinia umbrorum**
87	key point 10b (Ascospores biguttulate; apothe cia subsessile, cupulate, grayish tan, up) should be: 10b Ascospores eguttulate; apothe cia stipitate to substitpitate. . . 11
88	key point 4a **Scutellinia umbrarum** is: **Scutellinia umbrorum**
89	key point 7a **Meladina lechitina** is: **Miladina lechithina**
90	key point 19a (. . . . 19) should be: 20
	key point 20b add: or on mosses
91	key point 26a **Pyronema omphaloides** is: **Pyronema omphalodes**
101	**Scutellinia umbrarum** is: **Scutellinia umbrorum**

NAME CHANGES

Helvella acetabulum (L.ex St.Amans) Quel is:
    Paxina acetabulum (L.ex St.Amans) O. Kuntze
Helvella albipes Fuckel is:
    Helvella leucopus
Helvella elastica Bull.ex St.Amans is:
    Leptopodia elastica (Bull.ex St.Amans) Boud.
Helvella leucomelaena (Pers.ex Fr.) Karst. is:
    Paxina leucomelas (Pers.) O.Kuntze
Helvella macropus (Pers.) Nannf. is:
    Macroscyphus macropus Pers.ex S.F.Gray
Helvella villosa (Hedw.ex Kuntze) Dissing & Nannfeldt is:
    Cyathipodia villosa (Hedw.ex O.Kuntze) Boud.
Mitrula abietis Fr. is:
    Heyderia abietis (Fr.) Link
Mitrula gracilis is:
    Bryoglossum gracile (Karsten) Redhead
Paxina recurvum Snyder is:
    Gyromitra melaleucoides (Seaver) Pfister
Peziza melaleucoides Seaver is:
    Gyromitra melaleucoides (Seaver) Pfister
Plectania melastoma (Sow ex Fr.) Fuckel is:
    Plectania melastoma (Sow:Fr) Fuckel
Pseudoplectania melaena (Fr.) Boud. is:
    Pseudoplectania vogesiaca (Pers.) Seaver
Pseudoplectania nigrella (Pers.ex Fr.) Fuckel is:
    Pseudoplectania nigrella (Pers.:Fr.) Karsten
Sarcosoma mexicana (Ellis & Holway) Paden & Tylutki is:
    Plectania mexicana (Ellis & Howay in Holway) Paden
Sprageola irregularis is:
    Neolecta irregularis (Pk.) Redhead

1993

CONTENTS

PREFACE TO DISCOMYCETES

This work is the result of studies of Idaho Discomycetes during the past twenty years. Since the publication of "Some Edible and Poisonous Mushrooms of Idaho" in 1962 much new information has come to light to warrant an extensive revision of the limited treatment of common Idaho Discomycetes given at that time. The foundation of our understanding of Idaho Discomycetes was established by Paden (31) in 1967 in his dissertation on the Pezizales of Northern Idaho. This I found most useful in the preparation of this practical guide.

In view of my objective of presenting information on the conspicuous, the edible, the poisonous, or otherwise interesting fungi of Idaho, no attempt has been made to include all of the Discomycetes. Apart from the Geoglossaceae (Earth Tongues), I have excluded the inoperculate Discomycetes from this work because our studies of that group are incomplete and because many are quite small and inconspicuous and are therefore not of interest to the mushroom collector. In the operculate Discomycetes, on the other hand, are found many large and conspicuous fungi including some of the most edible and famous such as the morels and false morels. Furthermore, we know a great deal more about them. Consequently I have included 104 species of operculate Discomycetes known to occur in Idaho. However, descriptions and illustrations are presented only for those which can produce fruiting bodies that are 1 cm or larger in length or width. For the same reasons 12 inoperculate Discomycetes of the Geoglossaceae (Earth Tongues) are also included.

The collections used as a basis for this study were made by myself, my students and by many mushroom collectors throughout the state. To these I offer my sincerest thanks for the valuable information. Most collections were obtained from the forested regions extending from the Canadian border to the dry sage brush regions of the great basin of southern Idaho. Specimens are in the University of Idaho Herbarium. In addition, when the Idaho distribution records were incomplete, I consulted Larsen and Denison's (22) list of Pezizales of Northwest America.

The photographs were taken with a 35 mm Exakta camera with supplementary lenses using Panatomic X film or Kodachrome II color transparency film. These transparencies were then copied to black and white using a Kodak MP3 camera with polaroid back and a Bowens illumitran. The line drawings were prepared by Ann Curtis.

I thank the University of Idaho Stillinger Grant committee for financial support of part of this project and the Research Council and the Regents of the University of Idaho who initially supported this project as Special Research Project 68. My sincerest thanks to Dr. John Paden of the University of Victoria for valuable assistance and for reading and criticizing the manuscript. For continued interest and support I should like to thank Dr. Kenneth Laurence, Head, Department of Biological Sciences and finally I thank Pauly Waldron for typing the manuscript.

Mushroom habitat: A virgin cedar-hemlock forest on Granite Creek near Priest Lake in Northern Idaho. This is a world famous mushroom collecting area.

INTRODUCTION

Mushrooms of the group called Discomycetes are among the most conspicuous and brilliantly colored of any in that large group of fungi known as Ascomycetes. The unifying feature is the presence of a microscopic sac-like cell called the ascus, in which the ascospores are produced (Fig. 19c). Even without a microscope the general shape of the fruiting body will distinguish a discomycete from any basidiomycete type mushroom with which it may be confused. You will find the sponge-like cap of the morel, the saddle-shaped or brain-like caps of *Helvella* species, and the cup-shaped apothecia of a large number of Discomycetes are usually so unlike any Basidiomycetes that no confusion will arise. However, occasionally the fruiting bodies of some Discomycetes resemble those of certain Basidiomycetes and these could be misidentified if a microscope is not used. The earth tongues resemble certain coral fungi; the truffles resemble false puffballs; and very rarely some basidiomycetes are cup shaped. Fortunately these exceptions are not frequently encountered in our flora, and, when met with, can be handled readily by paying particular attention to the details provided in the field key. Of course, if after consulting the key a question still remains you will need to use a microscope to determine if asci or basidia are present. Other microscopic features are also very important.

Certain Discomycetes produce asci with a pore or irregular tear at the apex and are called inoperculate Discomycetes while others produce a lid or operculum, and are known as operculate Discomycetes. It is impossible to distinguish the two types on field characters alone. Therefore the use of a microscope is almost mandatory if you are faced with the identification of any rather small cup-shaped discomycete growing on wood. On the other hand, the soil inhabiting, large, conspicuous, noncupulate forms in both groups are readily distinguished on field characters.

When and Where to Collect Mushrooms

Finding mushrooms is not difficult. The foresters, hunters, fishermen, hikers, picnickers, and others who frequent the woods know that they can be found quite easily at certain times of the year. Mushrooms depend upon organic matter for nutrition and suitable temperature and abundant moisture for fruiting. These conditions are encountered in the cool moist forests and meadows in our mountainous areas. In the cities, good collecting can be had in well watered lawns, on golf courses, and in gardens, where occasionally morels may fruit on overwintered mulch.

In some parts of the state mushrooms are rare. The low elevation ponderosa pine-type forests are too dry and unproductive, except for short periods at certain times during the year. Few mushrooms, except certain puffballs, are found in the semi-desert Snake River plains and the dry rocky stretches along the Salmon and Clearwater Rivers. Sometimes, a change in elevation in these or other areas can increase the chance of success. In steep terrain, somewhere between the hot dry valleys below and the snows at higher elevations, an adequate supply of humus or organic matter exists and conditions for fruiting are satisfactory. Collecting there may be very good when few mushrooms are to be found elsewhere.

In general, best collecting of Discomycetes is encountered in the spring, shortly after the snows have disappeared. In the summer, collecting is fair at higher elevations where summer showers and melting snows provide moisture. After a heavy shower, it is best to wait about 3 days to a week before collecting. The most important factor determining our discomycete crop is the amount of precipitation received during the spring and early summer rainy periods. For specific information on when and where to collect consult the habitat-distribution section of the species of interest.

Distinguishing the Mushrooms from the Toadstools

The term "mushroom" refers to any of the large fleshy fungi technically called Ascomycetes and Basidiomycetes. These may be edible or poisonous and contrary to what you may have heard there are no tests which can distinguish the "toadstools" from the "mushrooms." One of the old favorites, the "silver spoon test," claims that poisonous mushrooms turn a silver spoon black. People have died by relying on it. Suffice it to say that this or any other test of this type has been shown to be unreliable. The term "toadstool," therefore, has no scientific meaning. The only way to distinguish the poisonous from the edible mushrooms is to know them by species, just as we distinguish oranges from lemons in the market. Anyone can do this by paying attention to certain details of the fruiting bodies. By species is meant a certain kind of mushroom which is different from other kinds or types. For example, we recognize different species or kinds of pine trees such as white pine, yellow pine, lodgepole pine, etc., all of which are different from each other, but yet they are similar enough to be recognized as belonging to the group called pines, or technically to the genus *Pinus*. So with the mushrooms, the different species are grouped into genera if they are sufficiently similar. Also, each mushroom species is designated by a binomial or scientific name consisting of two parts. The scientific name of the common morel or sponge mushroom is *Morchella esculenta*. The first mentioned part, *Morchella,* indicates the genus or generic name of the morel and the second part, *esculenta,* refers to a particular kind or species of morel, the Common Morel. A different morel would have the same generic name but a different specific epithet, e.g. *Morchella elata*, or Black Morel. The "species name" is the generic name plus the specific epithet. In most publications, the specific name is followed by an author citation. This is an abbreviation of the person or persons who first named the fungus or who made a new combination of old names.

Throughout the ages, man has accumulated records about toxic mushrooms from actual cases of poisoning backed by accurate identification of the species involved. We know that mushrooms produce a variety of effects in man and they do not affect every person in the same way. Some are deadly poisonous. Others are slightly poisonous, evoking a variety of disturbing symptoms; still others are disagreeable or tough and woody or otherwise indigestible; and lastly, some are edible and quite delicious. Not all the edible types are equally so, for some people may experience an adverse reaction, such as after eating eggs, cucumbers, strawberries, and the like. On the other hand, there are others that are wholesome and delicious and undisputedly edible. Anyone interested in an excellent account of mushroom poisonings is referred to the book by Lincoff and Mitchel (23).

The following precautions will aid in avoiding the poisonous ones and in reducing the chances of having an undesirable experience: (1) Be certain of your identification to the genus and species and that the mushroom is an edible one. You can blame no one but yourself for your mistakes. This publication will help you to identify some of the common edible species, but it does not include all species to be encountered. Therefore, if upon comparison, the unknown mushroom at hand does not fit the description and photographs, discard it unless you wish to study it for other purposes. (2) When eating a mushroom species for the first time, even the morel, try only a small amount for there is always the possibility of an adverse reaction. Each person should gain personal experience with each new species tried. (3) Learn to recognize the mushrooms that are responsible for serious poisonings, so that you can avoid them. (4) Do not eat mushrooms that are infested by larvae (as noted by small pin holes in the flesh most often starting in the base of the stalk), or are overmatured or discolored, or indicate in some other way an advanced stage of decay. Just because they are free for the taking does not mean one should not be selective in collecting for the table. (5) Be careful when eating immature stages of mushrooms, since species are hard to tell apart at this stage. You may have a primordium that is just as poisonous as a fully expanded easily identified specimen. (6) Avoid eating raw mushrooms, particularly the false morels and saddle fungi. Mushrooms should be thoroughly cooked before eating; and do not overeat for it may lead to a very worrisome case of ordinary indigestion. (7) If a poisoning occurs, call a doctor immediately. Induce vomiting with an emetic such as mustard in water or by tickling deep in the throat with the fingers. The mushroom eaten should be salvaged for identification for it may help in prescribing the best treatment.

Development and Parts of the Mushroom

In determining mushroom species, it is essential beforehand to know a few basic terms and facts concerning their structure and development. Mushrooms belong to the group of organisms known as fungi. In common with many other fungi, mushrooms are composed of two stages: the spawn or mycelium which is the somatic stage, and the fruiting bodies which represent the reproductive stage. The mycelium is found in the substrate, i.e. rotting stump, humus, duff, soil, decaying leaves, etc., where it obtains nutrients and water necessary for growth. Because it precedes the reproductive stage, any factors bearing on its growth will profoundly influence the appearance of the mushroom fruiting bodies. The mycelium is composed of very finely branched microscopic thread-like structures called hyphae. These, in the aggregate, are easily visible without a microscope. Mycelium can be found in any damp woods by merely turning over the top few inches of duff or decaying leaves and examining the white cottony threads coursing throughout. This mycelium (plural mycelia) arises from the spores, which are minute seed-like structures capable of growing into a new mushroom in a favorable spot. Some mycelia will produce fruiting bodies within a few weeks after growing from spores; others take longer periods, even years. Mycelia which are longlived remain dormant during periods of drought and severe cold or hot weather. When conditions of moisture and temperature are favorable, these mycelia may produce several crops of fruiting bodies, not only during the same season, but year after year as long as sufficient food is available. Knowing this, one can return to the same spot to collect a favorite edible species year after year. This is a good and safe practice. At the time of fruiting, the mycelium produces tiny knots that enlarge and expand rather rapidly giving the impression the mushroom arose overnight. Actually, the entire fruiting process takes quite some time and is often initiated by environmental factors. The main function of the fruiting body is to produce spores by which the mushroom is reproduced and perpetuated in nature.

When identifying Discomycetes it is the parts of the apothecium or fruiting body which are carefully studied. The chief parts of an apothecium are illustrated in Fig. 24 and are as follows: the hymenium, the margin, the exterior, the flesh and the stipe (if present). Those using a field approach should examine these with great care with the unaided eye or preferably with a 10X pocket lens. Do not be hesitant about cutting open the fruiting body to reveal the inner tissues. A microscopic study will reveal greater detail and show that the apothecium is composed of a hymenium of asci, ascospores and paraphyses, a subtending subhymenium or hypothecium and the excipulum which is usually partitioned into the inner medullary or ental excipulum and the outer ectal excipulum, see Figs. 25-27. You can learn the meaning of any unfamiliar descriptive terms used in the keys by consulting the glossary on p. 123 or the labeled diagrams of Plates I-III. Some apothecia are highly modified and complex. These exhibit a pileus-stipe arrangement with a variably shaped pileus on which the hymenium is born and a prominent stipe (Fig. 1-5). Others such as the truffles produce typically globose, and essentially closed ascocarps below the ground. The hymenium, if present, lines the inner wall cavities or convolutions (Fig. 17, 18).

Keys and List of Species

To aid you to determine the species of an unknown discomycete, I have prepared a number of keys. The **field key** to Idaho Discomycetes will be found on p. 13 and will be most useful to the beginning mushroom collector, whereas the **technical key** on p. 18 will require a greater knowledge of the subject. In the use of a key, simply compare the specimen at hand with the characteristics listed in the key. Starting with the first couplet descend through the key choosing each alternative to which the specimen applies. This is carried on until finally (as recognized by the binomial) the species is reached, or the name of a higher taxonomic category e.g. order, family or genus is obtained. If a specimen is identified to one of the higher taxonomic categories above the species level, then proceed to the key provided for that category and continue in a similar fashion until the binomial representing a species is obtained. Refer then to the description and photograph of that species for a verification of the identification. If considerable difficulty is encountered, an error in couplet choice may have been made or the specimen may be one not included here. For information on discomycetes not listed consult the technical literature listed in the references.

Enumerated under each family are the species of that group that occur in Idaho. Of these, I have elected to describe and illustrate only those which represent a type or which are large and common. Information on some species will be found under remarks given for closely related species.

In the descriptions, emphasis is on the characteristics such as size, color, and habitat etc., which cannot be observed readily in the photographs. The unknown specimen should be compared to both the text and photographs for a complete description. You will find the photographs very useful for determining such superficial characters as the shape of the fruiting body, length of stipe if present, wrinkling of the hymenium, hairiness of the margin or exterior surfaces and the general fruiting habit.

Whenever possible I used the color designation suggested in the ISCC-NBS color-name charts illustrated with centroid colors, supplement to National Bureau of Standards Circular 553. These color terms have the advantage of calling to mind an image of the color when the color charts on which these are based are not available.

The symbol *um* used in this publication refers to *micrometer* which is 10^{-6} meter or $1/25,000$ of an inch. The micrometer was formerly known as the "micron."

A composite photograph of choice and edible Discomycetes. Left to right: Pig's Ear, Black Morel, Walnut, and Brain Mushroom.

FIELD KEY TO IDAHO DISCOMYCETES

11a(10a)   Heads light brown to pinkish buff, conical to ovoid, smooth, stipe darker than head, growing on conifer needles . . . .
. . . . *Mitrula abietis* p. 114

11b(10a)   Heads ochraceous to orange-buff, cerebriform to rugose, stipe lighter than head, on ground in mosses . . . .
. . . . *Mitrula gracilis* p. 115

12a(10b)   Fruiting bodies drab or dark gray . . . . . *Cudonia grisea* p. 111

12b(10b)   Fruiting bodies cream to dark brown or pinkish buff 13

13a(12b)   Pileus pinkish buff to grayish brown, stipe glabrous . . . .
. . . . *Cudonia monticola* p. 111

13b(12b)   Pileus cream to dark brown, stipe furfuraceous, striate to ridged . . . . *Cudonia circinans* p. 111

14a(8b)   Fruiting bodies black to dark brown 15

14b(8b)   Fruiting bodies yellow to cinnamon buff . . . .
. . . . *Spathularia flavida* p. 115

15a(14a)   Fertile area velvety from setae   *Trichoglossum hirsutum* p. 116

15b(14a)   Fertile area smooth, not velvety from the presence of setae
. . . . *Geoglossum fallax, Geoglossum glabrum* and *Geoglossum nigritum* p.112
These are scarcely distinguishable on field characters. See *Geoglossum* in key to Geoglossaceae p. 109 for distinguishing microscopic characters.

16a(4b)   Fruiting bodies bright colored from the presence of carotenoid pigments: red, orange, bright yellow or bright yellowish brown 17

16b(4b)   Fruiting bodies somber colored, carotenoids absent: black, purple-black, violet, gray, brown, reddish brown, pale or dull yellowish brown 23

17a(16a)   Growing on wood or branches 18

17b(16a)   Growing on the ground, soil, dung, among moss, on leaf litter, duff, etc., not on wood 22

18a(16a)   Fruiting bodies orange to reddish orange, gelatinous, growing on dead (Fir) *Abies* branches that usually still retain many old needles; found in spring near or in snowbanks or shortly after the snows recede . . . . *Pithya vulgaris* p. 32

18b(16a)   Not on *Abies* branches near snowbanks 19

19a(18b)   Margin of fruiting body fringed with dark bristle-like hairs
. . . . 20

19b(18b)   Margin of fruiting body without hairs 21

20a(19a)   Fruiting bodies white to yellowish . . . .
. . . . *Humaria hemispherica* p. 104

20b(19a)   Fruiting bodies scarlet . . . .
. . . . *Scutellinia scutellata* and *Scutellinia umbrarum* p. 101

21a(19b)   Growing on burned wood 53

21b(19b)   Fruiting bodies on wet wood or submerged wood, surrounded by radiating white mycelium . . . .
. . . . *Meladina* p. 89 and *Psilopezia* p. 90

22a(17b)   Fruiting bodies small 1-2 mm wide pale yellowish to greenish, often darkening from maturation of dark colored spores,

34b(33a) Hymenium dark to black brown, exterior whitish or brown . 35

35a(34b) Apothecia large, up to 6cm wide, cupulate, sometimes flesh
exuding a yellowish juice when cut ....
.... *Plicaria endocarpoides* p. 58

35b(34b) Apothecia small, rarely exceeding 2 cm wide, discoid, flesh not
exuding a yellowish juice 52

36a(33b) Fruiting bodies small, up to 5 mm wide, soft and gelatinous,
pulvinate, reddish-brown........ *Pachyella babingtonii* p. 52

36b(33b) Fruiting bodies larger 1 cm or more in dia., fleshy, cupulate to
discoid or otidioid 37

37a(36b) Occurring on dung, or well manured soil.................
.... *Peziza vesciculosa* p. 56

37b(36b) On soil, duff, rotting wood, etc. not on dung.............38

38a(37b) Occurring on soil, near plaster, cement, etc. often in cellars,
caves, greenhouses, crawl spaces and the like...............
.... *Peziza domiciliana* p. 57

38b(37b) Occurring in woodlands, fields or in gardens associated with trees
....39

39a(38b) Fruiting bodies small, less than 2 cm wide, hymenium blackish
brown, on soil, summer .... *Peziza brunneoatra* p. 50

39b(38b) Fruiting bodies larger, dark brown to tan, on soil, moss or rotting
wood 42

40a(24b) Fruiting bodies deep cupulate to urn-shaped 51

40b(24b) Fruiting bodies shallow cupulate, discoid, turbinate, or cleft to
base on one side (otidioid)...........................41

41a(40b) Fruiting bodies turbinate, large, up to 8.5 cm in diameter, with
massive gelatinous interior, exterior black, wrinkled..........
.... *Sarcosoma mexicana* p. 29

41b(40b) Fruiting body cupulate, gel layer absent or very thin 49

42a(39b) Exterior of cup whitish, gray, tan, or pale yellowish-brown .43

42b(39b) Exterior of cup reddish brown or brown 44

43a(42a) Exterior conspicuously dotted with slightly darkened, coarse,
scales or pustules................... *Disciotis venosa* p. 35

43b(42a) Exterior not pustulate, or pustules, fine, and pale..........50

44a(42b) Fruiting bodies split on one side, the cleft extending to base or
nearly so, ear-shaped (otidioid)........ See *Otidea* p. 95-100

44b(42b) Fruiting bodies cupulate to discoid, if split not regularly one-
sided and extending to base...........................45

45a(44b) Fruiting bodies repand, occurring in early and mid-spring,
hymenium wrinkled or dimpled, yellowish to deep yellowish
brown, or strong brown 46

45b(44b) Fruiting bodies cupulate, occurring in late spring, summer and
fall, hymenium smooth, reddish brown 47

46a(45a) Fruiting bodies short stalked, hymenium yellowish brown to
strong brown, common ....
.........*Discina perlata* p. 68 and *Discina apiculatula* p. 66

46b(45a) Fruiting bodies sessile, hymenium yellow to bright yellowish
brown, rare..................... *Discina leucoxantha* p. 67

47a(45b) Occurring in late spring and summer ....
.... *Peziza badioconfusa* p. 49
47b(45b) Occurring in the autumn................ *Peziza badia* P. 48
48a(32b) Hymenium purple to purple brown, widely distributed in spring
on burned soil, old campfire sites, etc......................
................................*Peziza praetervisa* p. 52
48b(32b) Hymenium violet to reddish-violet, found in spring and summer
on heated soil and not as common as *Peziza praetervisa* ......
.... *Peziza violacea* p. 57
49a(41b) Fruiting bodies cleft on one side (otidioid) arising in a cluster,
fleshy .... *Wynnella silvicola* p. 62
49b(41b) Fruiting bodies cupulate, arising singly, gelatinous when young
.... *Sarcosoma latahenis* p. 29
50a(43b) Exterior of fruiting body marked with prominent branching ribs
(like a cabbage leaf) ....
Helvella acetabulum p. 74 and *Helvella grieseoalba* p. 74
50b(43b) Exterior without branching ridges................... *Peziza*
(see *Peziza repanda* p. 55, *Peziza varia* p. 55 and *Peziza sylvestris*
p. 54, three species which are scarcely separable on field
characters).
51a(40a) Fruiting body blackish-brown (smoky) whitish near white stipe,
which is ribbed, common, widespread ....
.... *Helvella leucomelaena* p. 78
51b(40a) Fruiting body exterior purplish-brown, hymenium pinkish
brown, whitish stipe not ribbed, rare, found in cedar-hemlock
zone only .... *Neournula pouchetii* p. 22
52a(35b) Hymenium blackish, exterior dark brown ....
.... *Plicaria trachycarpa* p. 59
52b(35b) Hymenium gray brown, exterior gray below, brownish toward
margin .... *Peziza petersii* p. 51
53a(21a) Fruiting bodies flat, clustered on a fluffy crust of mycelium
(subiculum)................... *Pyronema omphaloides* p. 91
53b(21a) Fruiting bodies deep cupulate, clustered or not but without
subiculum .... *Geopyxis carbonaria* p. 96

TAXONOMIC KEY TO IDAHO DISCOMYCETES

1a Fruiting bodies cup or disk-shaped, club-like and often with a flattened upper region (spatulate), or stalked with pileus; sponge-like (Morels); saddle-like; brain-like or globose (False Morels); or subterranean (Truffles); Spores borne in operculate or inoperculate asci .... DISCOMYCETES ... 2

1b Fruiting bodies fleshy, tough to woody or gelatinous; umbrella-like with tubes (pores), spines or gills on the underside, with or without cap and stalk arrangement, shelf-like; or ball-shaped, bird's-nest or star shaped, growing above or below ground; spores produced on septate or non-septate basidia; clamp connections often present .... BASIDIOMYCETES (To be treated in later works)

2a Fruiting bodies occurring under ground; ascospores typically disseminated by animals, not violently discharged .... TUBERALES, p. 105

2b Fruiting bodies on or above ground, if subterranean at first, then in part or wholly exposed at maturity; ascospores violently discharged 3

3a Saprobes usually on soil, duff, dung, etc., occasionally on well rotted softened wood; asci operculate, without apical pore; ascospores always one-celled, typically elliptic or globose, rarely elongate .... PEZIZALES, p. 19

3b Parasites or saprobes, typically on firm or hard wood, plant debris etc., and on soil; fruiting bodies clavate, spatulate, capitate or with pileus and stipe; asci inoperculate, with apical pore; ascospores one to many-celled, long and slender or curved, rarely globose or elliptic......... .... (Earth Tongues) GEOGLOSSACEAE, p. 108

THE PEZIZALES

The majority of species in this work are Pezizales. Fruiting bodies in the main are cup-shaped and vary in size from the minute (1 mm) to very large (up to 10 cm wide). Some, such as the morels and saddle fungi, produce conspicuous stalked fruiting bodies with variously shaped caps. Ascocarps are typically epigeous except for *Geopora cooperi* and *Sarcosphaera crassa* both of which initiate development below ground and only when mature do they become exposed and opened to release their spores. Except for *Rhizina undulata, Urnula craterium,* and *Caloscypha fulgens,* all species of Pezizales are saprobes, living on dead wood, litter, duff, soil, dung or humus.

Some species are edible and highly prized e.g. morels and bell morels while others are poisonous or suspected of being so e.g. *Gyromitra esculenta* and still others which include the vast majority of Discomycetes have not been tested. Best fruiting occurs during the spring and early summer but some may be found in late summer and fall. The group is divided into seven families of which the Morchellaceae and Helvellaceae are of greatest interest to those who collect for the table.

Diagnostic Description

Apothecia cupulate, pulvinate or stipitate-pileate, epigeous, or if hypogeous at least spores shot-off and aerially disseminated; asci operculate or suboperculate, typically cylindric; ascospores one-celled, globose to elliptical or fusoid, never septate or long needle-like, hyaline or pigmented, smooth or rough.

TAXONOMIC KEY TO FAMILIES OF PEZIZALES

1a Apothecia tough, mostly black or dark colored, at times quite gelatinous internally; asci not bluing in iodine, cylindrical with long hypha-like base, thick walled, with subterminal or terminal operculum, sometimes with interrupted ring at apex; excipular cells prosenchymatous; ascospores acyanophylic or with cyanophilic warts, symmetric to asymmetric, and with or without oil drops 2

1b Apothecia fleshy, fragile, never or rarely gelatinous within; asci thin-walled, bluing with iodine or not, typically with an apical operculum, without ring near apex, excipular cells prosenchymatous or parenchymatous or both 3

2a Apothecia dark-colored, brown to black; terrestrial or lignicolous . . . . 1. *Sarcosomataceae, p. 20*

2b Apothecia bright-colored, orange or scarlet, carotenoids present; on

19

firm wood only, typically on sticks and branches ....
.... 2. *Sarcoscyphaceae, p. 31*

3a. Ascocarps usually with cap and stalk, rarely discoid or cupulate; pileus sponge-like, wrinkled, or smooth and campanulate; carotenoids absent; ascospores eguttulate, 20-60 nucleate; external polar granules typically present in ascus .... 3. *Morchellaceae, p. 33*

3b Ascocarps sessile, or, if long stipitate, cap brain, saddle or cup-like, carotenoids present or absent; ascospores guttulate or eguttulate 1 to 4 nucleate; external polar granules absent 4

4a Ascospores thick walled when young, eguttulate, purple to purple brown at maturity, uninuculeate; usually on dung .. 4. *Ascobolaceae, p. 43*

4b Ascospores hyaline, if brown, pigment deposition occurring from within the spores, thin-walled when young, nuclear number variable; on various substrates 5

5a Asci in part of wholly bluing with iodine ...... 5. *Pezizaceae, p. 44*

5b Asci entirely non-bluing......................................6

6a Apothecia usually with prominent elongated stipe; carotenoids absent; pileus cupulate, discoid, or most often mitrate to gyromitroid; exterior usually not hairy (except in *H. macropus*); ascospores 1-3 guttulate, 4-nucleate .... 6. *Helvellaceae, p. 61*

6b Apothecia cupulate to discoid, sessile or short stalked but never pileate; carotenoids present or absent; margin hairy or not; ascospores uninucleate or rarely binucleate and varying from eguttulate to polyguttulate .... 7. *Pyronemataceae, p. 85*

1. Sarcosomataceae

This family will not be of interest to those who collect for the table for many of the species are quite small and none of the species have yet been tested for edibility. The fruiting bodies are typically dark colored, tough or gelatinous, and cupulate or deep cupulate. They appear most abundantly in the spring but extend their fruiting into summer. *Sarcosoma mexicana* with the copious gelatinous interior is the largest and most conspicuous member of the group.

Diagnostic Description
Apothecia dark, tough, at times very gelatinous, sessile or stipitate; asci long cylindric to clavate, thick-walled, not bluing in iodine, suboperculate; ascospores guttulate or not, hyaline, smooth or rough with cyanophilic warts, at times with a gelatinous sheath; paraphyses usually anastomosing, with uninucleate cells; medullary excipulum of textura intricata; on rotting wood, humus or soil, typically fruiting in the spring.

LIST OF SPECIES

1. *Neournula pouchetii* (Berthet & Riousset) Paden
2. *Plectania melastoma* (Sow ex Fr.) Fuckel
3. *Plectania milleri* Paden & Tylutki
4. *Plectania nannfeldtii* Korf
5. *Pseudoplectania melaena* (Fr.) Bond
6. *Pseudoplectania nigrella* (Pers. ex Fr.) Fuckel
7. *Sarcosoma latahensis* Paden & Tylutki
8. *Sarcosoma mexicana* (Ellis & Holway) Paden & Tylutki

KEY

1a   Spores elliptical, rough; apothecia deep cupulate to urceolate up to 4 cm wide and 5 cm high; hymenium pinkish brown; exterior purplish brown; stipe whitish and immersed in substrate; asci 280-400 x 12-15 um, suboperculate, maturing evenly, operculum eccentric; ascospores 23-32 x 8-10 um, eguttulate or with several small guttulae, smooth at first, warty at maturity; warts staining with heated cotton blue; on the ground in the duff, spring and summer, Cedar-Hemlock zone N. Idaho, and Pacific Northwest .... 1. *Neournula pouchetii*

1b   Spores ellipsoid to fusoid, smooth .... 2.

2a   Spores globose.. 7.

2b   Spores ellipsoid to fusoid .... 3.

3a   Apothecia sessile to short stipitate; excipulum gelatinous .... 4.

3b   Apothecia sessile to long stipitate; excipulum fleshy to tough, not gelatinous .... 5.

4a   Apothecia 2-4 cm wide; exterior grayish; hymenium purple then black; spores ellipsoid, with few small guttules at first then eguttulate, 9-12 x 24-37 um; hypothecium hyaline .... *Sarcosoma latahensis*

4b   Apothecia 6-8 cm wide, broad and gel filled; hymenium black; exterior black; spores ellipsoid to suballantoid, with 1-3 oil drops; hypothecium dark .... *Sarcosoma mexicana*

5a   Apothecia with long stipe, cupulate, black, appearing terrestrial but usually attached to buried wood; ascospores elliptical, smooth to faintly verucose, eguttulate when mature, 10-14 x 22-30 um; conifer forests early spring .... *Plectania nannfeldtii*

5b   Apothecia sessile or short stipitate, on wood but not usually buried .... 6.

6a   Apothecia with reddish orange granules on exterior; margin if split not stellately lobed; hymenium black; margin incurved; ascospores ellipsoid-fusoid, eguttulate, but with many small oil drops at first, 8-10 x 21-24 um; on rotting conifer wood ....
.... *Plectania melastoma*

6b   Exterior black; margin stellate; hymenium dark purple brown, black when dry; ascospores smooth, elliptical, eguttulate, 9-11 x 21-26 um; medullary excipulum penetrated by dark brown hyphae originating in ectal excipulum; on wood, June, Id., Or . . . . *Plectania milleri*

7a Apothecia short or sometimes long stipitate, up to 8 cm wide; hymenium light yellow brown at first, spotted black, then shiny black at maturity; exterior tomentum of sparse and short (up to 135 um) brown hairs; ascospores hyaline, 12-17 um, with granular contents; spring, on decaying conifer wood . . . . *Pseudoplectania melaena*

7b Apothecia sessile or short stipitate, up to 1.5 cm broad; hymenium black to dark-reddish brown; exterior with dense tomentum of long (up to 380 um) brown hairs; ascospores hyaline, with granular contents, 10-12 um; spring on the soil, at times burned, and on decaying conifer wood and litter . . . . *Pseudoplectania nigrella*

WESTERN URNULA
Neournula pouchetti (Bert. & Rious.) Paden

FRUITING BODIES: Gregarious; urceolate to deep cupulate, stipitate, margin incurved or straight, crenate, up to 3 cm wide; *hymenium* pinkish brown, smooth, *exterior* pale tan to pale purple brown; *flesh* tough, drying to leathery texture; *stipe* whitish up to 1 cm wide, immersed in substrate.

MICROCHARACTERS: *Spores* 23-32 x 8-10 um, oblong-elliptic, smooth, at first, warty at maturity, the warts cyanophilic, with or without numerous small oil drops. *Asci* 280-400 x 12-15 um, long cylindric, maturing unevenly, suboperculate, the opercula eccentric. *Paraphyses* filiform, branched, anastomosing, the tips encrusted with a pale brown amorphous substance, up to 4 um wide at apex, *Medullary excipulum* of textura intricata, the hyaline hyphae 3.5-5 um wide. *Ectal excipulum* of textura angularis, the inner cells hyaline, the outer cells with pale brown walls, up to 17 x 20 um and catenulate to form a sparse tomentum.

HABITAT-DISTRIBUTION: Rare, spring and summer, Pacific Northwest on cedar duff.

DESCRIPTIONS: refs. 31, 33.

REMARKS: Edibility untested. The fungus has an unusual distribution pattern. Paden and Tylutki (33) described *Neournula nordmanensis* from north Idaho material. This was later found to be the same as *Urnula pouchetii* described a few years earlier by Berthet and Riousset (1) from North Africa. The nearest relative is the eastern species *Urnula craterium*. The fungus is also known from southern Vancouver Island, B.C. and Eastern Canada.

ORANGE-BLACK ELF CUP
Plectania melastoma (Sow. ex Fr.) Fuckel

FRUITING BODIES: Single to gregarious or cespitose; cupulate, sessile or short stipitate, margin incurved often split but not lobed, up to 3 cm wide; *hymenium* black, smooth, glistening; *exterior* black, with a rusty tinge from the presence of orange granules encrusting the hairs especially near margin; *mycelium* black.

MICROCHARACTERS: *Spores* 21-24 x 8-10 um, ellipsoid-fusoid, smooth, with numerous oil drops at first, eguttulate at maturity. *Asci* 380-450 x 11-13 um, suboperculate. *Paraphyses* brown filiform, branched, up to 4 um wide. *Hypothecium* of textura angularis, sharply delimited from the medullary excipulum and often pulling

23

away. *Medullary excipulum* of textura intricata in a gelatinous matrix, the hyphae 2-4 um wide. *Ectal excipulum* of inner hyaline and outer dark brown textura angularis, the outer cells forming a twisted olive brown tomentum encrusted with orange granules which dissolve in KOH to give a violet solution.

HABITAT-DISTRIBUTION: Widespread on conifer litter, spring. U.S. and Europe.

DESCRIPTIONS: refs. 6, 31, 37.

REMARKS: Edibility untested. A rather distinctive fungus readily distinguishable by the orange black exterior. The margin never splits stellately as in *P. milleri.*

MILLER'S BLACK ELF CUP
Plectania milleri Paden & Tylutki

FRUITING BODIES: Gregarious; cupulate to discoid, sessile or short stipitate, margin stellate, up to 4 cm wide; *hymenium* dark brown to purple brown, black when dry; *exterior* black, hairy.

MICROCHARACTERS: Spores 21-26 x 9-11 um, elliptical, smooth, hyaline, eguttulate. *Asci* 380-400 x 12-15 um, long cylindric, suboperculate. *Paraphyses* filiform, branched occasionally, up to 3 um wide. *Hypothecium* a dark zone of textura intricata. *Medullary*

excipulum of hyaline textura intricata in a gelatinous matrix traversed by dark branched septate hyphae that originate in ectal excipulum and terminate below the hypothecium. *Ectal excipulum* of dark celled textura angularis giving rise to the tomentum.

HABITAT-DISTRIBUTION: On duff in conifer woods late spring, Idaho and Oregon, rare.

DESCRIPTIONS: refs. 31, 33.

REMARKS: The stellate margin which might indicate a relationship to *Urnula* is diagnostic. Its closest relative is *P. melastoma*. Edibility untested.

NANNFELDT'S PLECTANIA
Plectania nannfeldtii Korf

FRUITING BODIES: Single to gregarious; cupulate, long stipitate, margin curved inward, up to 1.5 cm wide; *hymenium* black, smooth; *exterior* black, with a delicate tomentum; *stipe* long cylindric, narrowing toward the base which is usually attached to buried wood; up to 6 cm long; *mycelium* blackish.

MICROCHARACTERS: Spores 22-30 x 10-14 um, elliptical, smooth or faintly verrucose, eguttulate at maturity. *Asci* 350-480 x 11-14 um, suboperculate, J-. *Paraphyses* filiform, anastomosing, the tips embedded in a brown amorphous matrix, up to 4 um wide. *Medullary excipulum* up to 1 mm thick, of hyaline textura intricata in a thin gelatinous matrix, the hyphae 3-14 um wide. *Ectal excipulum* of textura angularis the inner cells hyaline the outer dark brown and giving rise to a tomentum of dark, smooth, twisted, septate hyphae.

HABITAT-DISTRIBUTION: Common in the Rocky Mountains and Pacific Coast states in the spring, shortly after the snows recede. Typically on buried *Picea* or *Abies* litter.

DESCRIPTIONS: refs. 31, 37.

REMARKS: Edibility untested. The apothecia are usually attached to buried sticks by a long stipe. The collector who does not dig down to retrieve the base will miss this point.

DARKENING FALSE PLECTANIA
Pseudoplectania melaena (Fr.) Boud.

FRUITING BODIES: Gregarious to cespitose; cupulate to repand, short to long stipitate, margin often wavy due to mutual pressure of

crowded fruiting bodies, up to 8 cm wide; *hymenium* yellow-brown to shining black at maturity, smooth at first, then wrinkled or umbilicate; *exterior* black, with short brown hairs.

MICROCHARACTERS: *Spores* 12-17 um wide, spherical, smooth, hyaline, contents granular. *Asci* 300-350 x 9-11 um, long cylindric, suboperculate. *Paraphyses* filiform branched mostly below midpoint, occasionally anastomosing, obscurely septate, often with small protuberances near tip, apices up to 3 um wide. *Hypothecium* of hyaline densely interwoven textura intricata, the hyphae 3-5 um wide. *Medullary excipulum* of hyaline textura intricata in a gelatinous matrix, the hyphae 3-7 um wide. *Ectal excipulum* of inner hyaline and outer dark brown textura angularis, giving rise to the stiff tomentum hyphae.

HABITAT-DISTRIBUTION: Widespread on decaying conifer wood, U.S. and Europe, spring, common.

DESCRIPTIONS: refs. 31, 37.

REMARKS: Edibility untested. The color of the hymenium is variable especially in young specimens. The yellow brown color persists for varying periods but later the apothecium is mottled with black and then finally glistening black. The stipe length also varies and at times may approach the length of *Plectania nannfeldtii*.

Mushroom habitat: Stanley Basin, an excellent mushroom collecting area in south-central Idaho. Morels may be found here in July.

BLACK FALSE PLECTANIA
Pseudoplectania nigrella (Pers. ex Fr.) Fuckel

FRUITING BODIES: Scattered to gregarious; globose at first then expanding to shallow cupulate and discoid, sessile to short stipitate, margin slightly incurved and often wavy, stiff, hairy, up to 1.5 cm wide; *hymenium* black to dark reddish brown, smooth; *exterior* black, hairy, especially at the base.

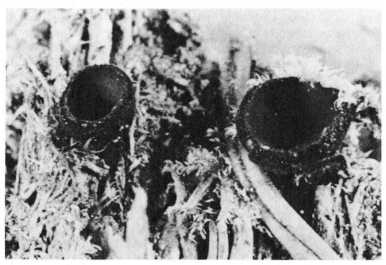

MICROCHARACTERS: *Spores* 10-12 um in diameter, globose, smooth, hyaline with granular contents. *Asci* 250-325 x 12-17 um, cylindric, with a long base, suboperculate. *Paraphyses* filiform, branched, often branched near tips, smooth and at times imbedded in an amorphous brown substance, up to 4 um wide. *Hypothecium* of densely interwoven, hyaline, textura intricata, the hyphae up to 4 um wide. *Medullary excipulum* of hyaline textura intricata, the hyphae up to 7 um wide. *Ectal excipulum* of hyaline inner and dark outer textura angularis, the cells up to 15 um wide. The marginal cells give rise to the dark tomentum hyphae and marginal bristles.

HABITAT-DISTRIBUTION: Widespread throughout the U.S. on rotting coniferous wood, litter and soil, spring and summer, common.

DESCRIPTIONS: refs. 6, 31, 37.

REMARKS: Edibility untested. This fungus could be confused with other black cup fungi, but the globose spores and the long tomentum hyphae at the point of attachment readily distinguish it.

LATAH SARCOSOMA
Sarcosoma latahensis Paden & Tylutki

FRUITING BODIES: Singles to gregarious; turbinate at first then discoid substipitate, texture gelatinous at first, becoming less so at maturity, up to 4 cm wide; *hymenium* purple black to black, undulate; *exterior* grayish to black, tomentose.

MICROCHARACTERS: Spores 24-37 x 9-12 um, elliptical, smooth, with several oil drops at first, eguttulate at maturity. *Asci* 380-475 x 10-14 um, suboperculate, long cylindric, gradually tapering toward base. *Paraphyses* filiform or slightly clavate, branching, anastomosing, septate, the tips up to 5 um wide and imbedded in amorphous olive-brown substance. *Hypothecium* of hyaline textura intricata approximately 50 um thick, the hyphae 3-4 um wide. *Medullary excipulum* of hyaline textura intricata in a gelatinous matrix, the hyphae 5-10 um wide. *Ectal excipulum* of hyaline inner and dark brown textura angularis, tomentum hyphae smooth, branched, septate, sometimes with globoid swellings up to 8 um wide and a 400 um long, tapering slightly and usually paler at the distal end.

HABITAT-DISTRIBUTION: On wood, litter and soil often near snowbanks, in coniferous forests of the Pacific Northwest, spring.

DESCRIPTIONS: refs. 31, 34.

REMARKS: Edibility untested. The amount of gel in mature fruiting bodies is quite variable and may be dependent upon environmental conditions.

MEXICAN GEL-CUP
Sarcosoma mexicana (Ellis & Holloway)
Paden & Tylutki

FRUITING BODIES: Single to gregarious; turbinate then discoid, up to 9 cm wide, filled with a thick gelatinous mass; *hymenium* black, concave to flat, margin raised; *exterior* black, usually wrinkled, covered with tomentum of dark hyphae.

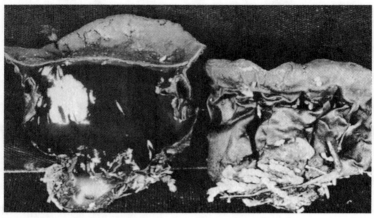

MICROCHARACTERS: *Spores* 23-34 x 10-14 um, ellipsoid, to suballantoid, smooth, hyaline with 1-3 oil drops. *Asci* 500-520 x 12-15 um, long cylindric, gradually tapering to base, suboperculate. *Paraphyses* filiform to slightly clavate, branched, anastomosing, closely septate, tips rounded or irregularly lobed up to 5 um wide, embedded in dark green amorphous material. *Hypothecium* a dark green zone of densely interwoven textura intricata, the hyphae 3-5 um wide. *Medullary excipulum* of textura intricata in a massive gelatinous matrix, the hyphae 3-5 um wide, hyaline. *Ectal excipulum* of dark textura angularis, tomentum hyphae dark green or brown, irregular, often roughened with dark granules, up to 6 um wide.

HABITAT-DISTRIBUTION: Widely distributed, but rarely abundant in western North America, spring, summer, fall on decaying wood or duff.

DESCRIPTIONS: refs. 31, 33, 40.

REMARKS: Edibility untested. A rather striking species when cut open to show the massive gelatinous zone. No other western species approaches it in this regard. In *Sarcosoma latahensis* the gel layer is reduced as the fruiting body matures, and is further distinguished by the characters emphasized in the key.

2. Sarcoscyphaceae

Pithya is the sole representative of this family in this area. *Sarcoscypha coccinea* also of this family is mentioned because I suspect it may occur in Idaho, but as of this date, I have no record of it. This family will be of no interest to those who collect for the table because the species are too small and of unknown edibility.

Diagnostic Description
Apothecia sessile to short stipitate, tough, hymenium, bright red to orange; asci not bluing, thick-walled, cylindrical, suboperculate; ascospores smooth, hyaline, thin-walled, globose or elliptical; paraphyses of multinucleate cells; medullary cells elongated; occurring in very early spring, strictly on wood.

LIST OF SPECIES
1. *Pithya cupressina* (Batch ex Fr.) Fuckel
2. *Pithya vulgaris* Fuckel
3. *Sarcoscypha coccinea* (Fr.) Lambotte

KEY

1a Apothecia small, 3-10 mm wide, yellow to orange, discoid or shallow cupulate, somewhat gelatinous when fresh, sessile to substipitate; ascospores smooth, hyaline, 10-12 um, globose; paraphyses with pale yellow droplets; saprophytic on dead *Abies* branches that usually still retain many old needles; early spring, at times in snowbanks; Idaho, Wa., Or., Ca. . . . . *Pithya vulgaris*

1b Apothecia large, 1.5-3 cm wide, scarlet, deep cupulate, usually stipitate ascospores smooth, hyaline, 10-12 x 26-40 um long, ellipsoid, with 2 large and numerous small globules; paraphyses with red granular contents; saprophytic on partially buried hardwood sticks; Oregon, Washington, west of Cascades, California and eastern U.S., early spring . . . . *Sarcoscypha coccinea* (see notes under *Pithya*)

COMMON PITHYA
Pithya vulgaris Fuckel

FRUITING BODIES: Gregarious; convex often umbilicate, sessile to substipitate up to 1 cm in diameter; *hymenium* orange to orange red, but pale yellow at first, especially if found under the snow; *exterior* paler to whitish, without hairs. *Mycelium* white.

MICROCHARACTERS: Spores 12-14 um in diameter, spherical, smooth, hyaline with granular contents at first then with a single large oil drop. *Asci* 300-340 x 12-14 um, long cylindric, suboperculate. *Paraphyses* slightly clavate, branched, occasionally anastomosing, tips up to 4 um wide, filled with pale yellow droplets. *Hypothecium* thin and scarcely differentiated. *Medullary excipulum* conspicuous at times separable, of textura globulosa and textura angularis, the outer cells giving rise to a sparce tomentum.

HABITAT-DISTRIBUTION: Common in early spring throughout the Pacific Northwest, on solid decaying *Abies* branches which usually still retain their needles. The habit of growing on recently dropped branches suggests it may be a parasite which infects the branch earlier and forms fruitbodies later after the branch overwinters on the ground.

DESCRIPTIONS: refs. 6, 31, 37.

Pithya cupressina

REMARKS: Edibility unknown. This is the only orange cup fungus found on *Abies* branches. It typically develops fruitbodies under the snow. These do not mature until the snows recede. *Pithya cupressina* occurs on cedar and is not gelatinous. *Sarcoscypha coccinea* the only other relative of this family in the Pacific Northwest is bright red and grows on hardwood sticks. The fungus is rare in the west but common in the eastern U.S. hardwood forests. It has not yet been found in Idaho, but it may rarely occur here.

3. Morchellaceae

This family contains the morels which are some of the most famous, delicious and distinctive mushrooms known. All members of this group are edible, but it should be noted that some morels produce adverse effects in some people. Apart from *Disciotis venosa* which is cupulate, all morels develop a fruiting body with a prominent hollow stalk and a wrinkled to pitted cap. They fruit on the ground in the spring but at higher elevations they may be found later in the year. Rarely morels will fruit in the fall. The external shape of morels varies considerably and this has led to considerable confusion in the classification of this group. I have taken a conservative view because it is impossible to utilize current species concepts based on gross morphological features which vary so much in our western material.

Diagnostic Description
Ascocarps fleshy, pileate, morchelloid, verpoid or discoid (in *Disciotis* only), typically brown to gray, never bright colored from carotenoids; asci clavate to cylindric, thin walled, not bluing with iodine, operculate; ascospores hyaline, with adhering exterior granules when fresh, eguttulate within, 20-60 nucleate, thin-walled at first; paraphyses not anastomosing; on humus and soil; spring.

LIST OF SPECIES
1. *Disciotis venosa* (Pers. ex Fr.) Boudier
2. *Morchella angusticeps* Pk.
3. *Morchella crassipes* (Vent.) Pers.
4. *Morchella elata* Fr.
5. *Morchella esculenta* Pers.
6. *Morchella semilibera* Fr.
7. *Ptychoverpa bohemica* (Krombh.) Boud.
8. *Verpa conica* Swartz ex Fr.

33

KEY

1a Pileus attached to stipe at margin or free up to mid-point only, surface honeycombed 2

1b Pileus discoid or cupulate or not pitted in honeycomb fashion, smooth or if wrinkled with vein-like folds then attached to stipe at apex only . . 4

2a Pileus margin free from stipe one-third to one-half the distance to apex; ridges whitish, not or slightly darkening; spores 22-28 x 12-16 um, ellipsoid, hyaline; in duff; at edges of woods . . . . *Morchella semilibera*

2b Pileus margin attached to stipe or nearly so 3

3a Pileus usually conic; ribs tan to gray, becoming smoky brown to black, in age the pits paler and longitudinally arranged; ascospores 24-26 x 12 x 14 um, ellipsoid, hyaline; paraphyses enlarged; on duff . . . .
. . . . *Morchella elata* and *Morchella angusticeps*

3b Pileus usually more elliptical or subglobose; ribs tan or grayish yellow, not darkening to smoky brown; pits irregularly arranged; ascospores 20-25 x 12-14 um, ellipsoid, hyaline; paraphyses enlarged; on conifer duff, spring, with peak fruiting occurring after *M. elata* has fruited
. . . . *Morchella esculenta* and *Morchella crassipes*

4a Stipe short or absent; pileus cupulate or discoid, up to 20 cm wide; hymenium folded or wrinkled vein-like, brown; exterior whitish with pustules; ascospores 21-24 x 12-14 um, smooth, elliptical, eguttulate but often with external polar granules; on the ground in conifer woods
. . . . *Disciotis venosa*

4b Stipe prominent, hollow; pileus campanulate, pendant and attached only at stipe apex; hymenium smooth or wrinkled 5

5a Pileus smooth, brown; stipe white; ascospores 28-32 x 15-19 um, eight per ascus, elliptical, smooth, hyaline, eguttulate, with external polar granules; spring; on soil . . . . *Verpa conica*

5b Pileus with longitudinal folds or anastomosing ribs, tan to yellow brown; stipe white; ascospores 54-66 x 17-18 um, two per ascus, ellipsoid-fusoid, smooth, hyaline, eguttulate, with external polar granules; on soil; spring . . . . *Ptychoverpa bohemica*

VEINY CUP FUNGUS
Disciotis venosa (Pers. ex Fr.) Boud.

FRUITING BODIES: Single to gregarious; cupulate to discoid, sessile or short stipitate, margin often splitting, flesh brittle, up to 20 cm wide; *hymenium* brown veined or convolute; *exterior* whitish, pustulate.

MICROCHARACTERS: *Spores* 21-24 x 12-14 um, elliptical, smooth, eguttulate with external polar granules. *Asci* 370-400 x 18-20 um. *Paraphyses* gradually clavate, stout, tips up to 12 um wide. *Medullary excipulum* of loosely interwoven textura intricata.
HABITAT-DISTRIBUTION: On soil or duff under conifers, widespread but not very common. Spring.
DESCRIPTIONS: refs. 6, 31, 37.
REMARKS: Edible and delicious. The eguttulate spores with external polar granules allies this typical cup fungus with the morels. On the basis of the cupulate fruit body and somber colors it could easily be mistaken for one of the Pezizaceae.

NARROW-CAP MOREL
Morchella angusticeps Pk.

FRUITING BODIES: Single to gregarious; *pileus* narrowly conic, grayish at first then blackening especially on the ribs, up to 5 cm wide, pits elongating; *stipe* cream to flesh colored, with delicate warts, about as wide as the pileus and up to 20 cm long, hollow.

MICROCHARACTERS: *Spores* 24-28 x 12-14 um, ellipsoid, hyaline, but yellowish in spore print. *Asci* 250-280 x 18-20 um. *Paraphyses* clavate, stout, the tips up to 17 um wide. *Medullary excipulum* of textura intricata. *Ectal excipulum* of textura globulosa, outer cells catenulate and cohering to form small warts. HABITAT-DISTRIBUTION: Widespread in conifer and hardwood forests, common in the spring. Often found in burns along with *M. elata*.

DESCRIPTIONS: refs. 37, 39.

REMARKS: Edible but see remarks under *M. elata*. This fungus intergrades with *M. elata* and may simply represent a growth form of that species.

BLACK MOREL
Morchella elata Fr.

FRUITING BODIES: Scattered; *pileus* conic, but very variable in size and shape, yellow brown with dark or smoky ridges or entirely darkening, ridges or pits longitudinally aligned, margin attached or slightly free, up to 4 cm wide and typically wider than the diameter of the stipe; *stipe* cream, rough inside and out, up to 10 cm long.

MICROCHARACTERS: *Spores* 24-28 x 12-14 um ellipsoid, smooth. *Asci* 300 x 20 um. *Paraphyses* clavate, stout, tips up to 17 um wide. *Medullary excipulum* and *ectal excipulum* as in *M. angusticeps*.

HABITAT-DISTRIBUTION: On the ground in conifer woods throughout the Pacific Northwest in the spring and rarely in the fall. It is to be expected later at higher elevations. Our most common morel. Sometimes it fruits explosively in the spring the year following a forest fire. It also occurs in gardens and the like where bark or woody mulch has been applied.

DESCRIPTIONS: refs. 6, 31.

REMARKS: A very variable species which intergrades with *Morchella semilibera* and *M. angusticeps*. Other than gross morphology no consistent features have been found to separate these. Edible to some, poisonous to others, evoking gastrointestinal upset. Many people first try this species when eating wild mushrooms and it is possible consumption of old or decayed worm infested specimens may be the cause. In informally surveying mushroom clubs in the Pacific Northwest, I find approximately one percent of the people have had adverse reactions from this species. It is also known as *M. conica*.

SPONGE MUSHROOM
Morchella esculenta Pers.

FRUITING BODIES: Scattered; *pileus* elliptic to subglobose, tan to yellow or grayish brown, ridges and pits irregularly arranged, or aligned radially, much broader than stipe, up to 8 cm wide; surface hoary or granulose rough; *stipe* whitish to pale yellow, hollow, slightly larger at base, longitudinally depressed in places, surface dry and granulose rough.

MICROCHARACTERS: *Spores* 20-25 x 12-14 um, ellipsoid, smooth, hyaline, eguttulate, *Asci* 280 x 20 um, J-. *Paraphyses* clavate the tips up to 15 um wide. *Medullary excipulum* and *ectal excipulum* as in *M. angusticeps*.

HABITAT-DISTRIBUTION: On the ground in conifer woods throughout the Pacific Northwest, sometimes in orchards. It fruits a little later or at about the same time as the black morel which at lower elevations in Idaho is May or June. Not as commonly encountered as the black morel.

DESCRIPTIONS: refs. 6, 31, 37.

REMARKS: Edible and choice. Reports of morel poisoning are usually traced to the black morel. *M. crassipes* is presumably the same except for the enlarged stipe base.

Sponge Mushroom *(Morchella esculenta)*

HALF-FREE MOREL
Morchella semilibera C.D. ex Fr.

FRUITING BODIES: Scattered; *pileus* subconic, pale to medium yellow brown, the ridges darkening slightly and pits elongating, margin free one-third to one-half the distance to apex; *stipe* whitish, granulose, hollow, short at first finally up to 8 cm long, base larger than apex.

MICROCHARACTERS: *Spores* 22-28 x 12-16 um, ellipsoid, smooth, hyaline, eguttulate. *Asci* 300 x 25 um. *Paraphyses* slightly clavate, septate, up to 18 um wide. *Medullary excipulum* of textura intricata. *Ectal excipulum* of textura globulosa.
HABITAT-DISTRIBUTION: On the ground in conifer, cottonwood and alder stands, widespread in early spring.
DESCRIPTIONS: refs. 6, 37, 39.
REMARKS: The half free morel is very close to and intergrades with the black morel. For edibility see remarks under black morel. This fungus is also known as *Mitrophora semilibera*.

EARLY BELL MOREL
Ptychoverpa bohemica (Kromb.) Boud.

FRUITING BODIES: Scattered; *pileus* yellowish brown, bell-shaped, attached only at apex of stalk, margin free, surface ridged and reticulated, up to 3 cm high and 2 cm wide; *stipe* whitish, tapering upward, stuffed then hollow, smooth to slightly floccose, up to 13 cm high.

MICROCHARACTERS: *Spores* 54-66 x 17-18 um, elliptic-fusoid, smooth, eguttulate with external polar granules. *Asci* 300-350 x 20-22 um, cylindrical, two spored. *Paraphyses* slightly clavate, stout; the apices up to 10 um wide.
HABITAT-DISTRIBUTION: On the ground under shrubs along ravines, river bottoms and mountain sides. Widely distributed, common during early spring.
DESCRIPTIONS: refs. 31, 37.
REMARKS: Edibility questionable. Some people who eat this fungus suffer gastrointestinal upset and lack of muscular coordination. Others have no difficulty. It is possible a gyromitrin like toxin may be produced by this fungus, therefore, parboiling is recommended.

BELL MOREL
Verpa conica Swartz ex Fr.

FRUITING BODIES: Single to scattered; *pileus* bell-shaped or subconic, brown, smooth or with few faint furrows near margin, white underneath, up to 2-3 cm wide; *stipe* white, tapering upward, smooth or slightly floccose giving a transverse striate appearance, hollow but loosely stuffed with hyphae.

MICROCHARACTERS: *Spores* 28-34 x 15-19 um, elliptical, smooth, hyaline, eguttulate with external polar granules. *Asci* 500-550 x 21-27 um, eight spored. *Paraphyses* gradually clavate, branched, septate up to 10 um wide at apex. *Medullary excipulum* of loosely interwoven textura intricata.

HABITAT-DISTRIBUTION: On the ground associated with hardwoods and grasses in river valleys, widespread throughout the U.S. in the spring, not common.

DESCRIPTIONS: refs. 6, 31, 37.

REMARKS: Edible but seldom encountered in any quantity. The cap is not always sharply conic as the specific epithet would imply.

4. Ascobolaceae

The fungi in this family produce small apothecia usually on dung and may be obtained readily by placing dung in culture vessels in the laboratory. Further information and keys for the determination of these microfungi may be obtained in the following references: 2, 18, 19, 36, 38. In view of the objectives of this publication illustrations and descriptions of Ascobolaceae are not provided. None are considered to be edible.

Diagnostic Description
Apothecia fleshy, pulvinate to discoid, usually small, sessile, with or without carotenoids, sometimes darkening when spores mature; asci operculate not bluing or bluing only in part, cylindrical to clavate; ascospores hyaline, purple or brown, smooth, striate, or verrucose, ellipsoid, uninucleate; usually on dung.

LIST OF SPECIES
1. *Ascobolus crenulatus* Karst.
2. *Ascobolus furfuraceus* Pers.
3. *Ascobolus geophilus* Seaver
4. *Iodophanous carneus* (Pers.) Korf.
5. *Saccobolus depauperatus* (Berk. & Br.) Hansen
6. *Saccobolus versicolor* (Karst.) Karst.
7. *Thecotheus cinereus* (Cr. & Cr.) Chen.

KEY
1a Ascospores united into a compact cluster........................2
1b Ascospores free from one another 3
2a Apothecia sessile, seldom over 300 um wide, convex, yellow then brown; asci protruding, bluing with iodine; ascospores ellipsoid-fusoid, smooth, purple brown at maturity, 6-8 x 11-15 um; spore cluster 10-16 x 30-37 um; paraphyses simple, 3 um wide at tip; on dung................
....*Saccobolus depauperatus*
2b Apothecia 0.5 - 2 um wide, pulvinate, pale violet darker in age; asci dark protruding, bluing with iodine; ascospores ellipsoid, purplish brown, 13-21 x 6-9 um, smooth at first finely warted or reticulate at maturity; cluster 40-60 x 14-20 um; paraphyses branched, 7 um wide at tip; on dung..
....*Saccobolus versicolor*
3a Ascospores purple or purple brown at maturity 4
3b Ascospores permanently hyaline; asci bluing with iodine 6
4a On soil; ascospores coursely granular, or reticulated or warty; apothecia sessile, up to 5 mm wide, discoid, greenish yellow at first, brownish in age; asci not bluing; spores ellipsoid, maturing from hyaline to purple and

finally purplish-brown, 9-13 x 19-23 um; paraphyses 2 um wide at apex, filiform or nearly so . . . . *Ascobolus geophilus*

4b On dung; ascospores striate 5

5a Ascospores large, 16 um or larger; apothecia up to 5 mm wide, yellowish to olive green or brownish with dark asci protruding; exterior furfuraceous; margin denticulate; asci bluing with iodine; ascospores 10-14 x 18-27 um, ellipsoid, with longitudinal anastomosing striae; paraphyses typically filiform 2-4 um wide; on dung . . . . . . *Ascobolus furfuraceus*

5b Ascospores smaller, 6-8 x 9-15 um; apothecia up to 2 mm wide, greenish yellow with dark asci protruding; exterior furfuraceous or granular; margin crenulate; asci bluing with iodine; spores ellipsoid, with longitudinal striae, 6-8 x 9-15 um; paraphyses 8 um wide at the tip; on dung . . . . *Ascobolus crenulatus*

6a Apothecia orange to flesh colored, pulvinate, 1-1.5 mm wide; asci bluing with iodine; ascospores elliptical, minutely verrucose, eguttulate, hyaline, 10-12 x 17-22 um; paraphyses 8 um wide at tip, stout; on dung . . . . *Iodophanous carneus*

6b Apothecia discoid, purple, 1-1.5 mm wide; asci bluing with iodine; ascospores ellipsoid, with minute cyanophilous warts, apiculate, hyaline, eguttulate, with gelatinous sheath at maturity, 13.5 - 15.5 x 31.5 - 40 um; paraphyses 4-6 um wide at apex; on cow dung . . *Thecotheus cinereus*

5. Pezizaceae

This is a large family well represented in Idaho, yet the mushroom collector who is seeking edibles will find this group a disappointment. Not only are they essentially untested as to edibility, but *Sarcosphaera crassa* is said to be poisonous to some people (23). Those who do not use a microscope will find it impossible to recognize the family and to identify the genera and species. The presence of asci which turn blue (J+) when mounted in Melzer's reagent and the generally cupulate habit distinguishes the group from all other operculate Discomycetes. Members of this family lack the brilliant red and orange colors found in the fruiting bodies of many Pyronemataceae.

Diagnostic Description
Apothecia fleshy to brittle, neither tough nor gelatinous, cupulate to discoid, sessile or stalked but not pileate; asci thin walled, clavate to cylindric, with apical operculum, bluing partly or rarely entirely with iodine (J+); ascospores thin-walled, hyaline or brown, uninucleate, spherical or elliptical; paraphyses not anastomosing.

LIST OF SPECIES

1. *Pachyella babingtonii* (Berk.) Boud.
2. *Peziza ammophila* Durieu & Montagne
3. *Peziza badia* Pers. ex. Merat
4. *Peziza badioconfusa* Korf
5. *Peziza brunneoatra* Desm.
6. *Peziza domiciliana* Cooke
7. *Peziza echinospora* Karst.
8. *Peziza petersii* Berk. & Curt.
9. *Peziza praetervisa* Bres.
10. *Peziza repanda* Pers. ex. Fr.
11. *Peziza sylvestris* (Boud.) Sacc. & Trott.
12. *Peziza varia* (Hedw.) Fr.
13. *Peziza vesiculosa* Bull. ex. Fr.
14. *Peziza violacea* Pers.
15. *Plicaria endocarpoides* (Berk.) Rifai
16. *Plicaria trachycarpa* (Curr.) Boud.
17. *Sarcosphaera crassa* (Santi ex Steudl.) Pouzar

KEY

1a Apothecia hypogeous or nearly so, opening in a stellate manner; in duff or sand 2

1b Apothecia epigeous, on ground or on wood; margin entire, or if splitting, then not splitting stellately 3

2a Apothecium at first cup-shaped, margin splitting finally, then flattening; hymenium brown; stipe remaining immersed in sand; spores smooth, elliptical, 15-10 um; paraphyses straight, up to 7 um at tip . ... *Peziza ammophila*

2b Semi-immersed in conifer duff; a hollow sphere at first, then cup-shaped with stellate margin; hymenium purple, violet, or tan-violet; ascospores 1-3 guttulate, smooth or minutely verrucose, elliptical, 7-9 x 15-22 um; paraphyses with purple granules, occasionally branched; spring, widespread .... *Sarcosphaera crassa*

3a Apothecia pulvinate, sessile, soft and gelatinous, small, 1-5 mm wide, reddish brown, transluscent; asci bluing throughout; ascospores elliptical, smooth, hyaline, biguttulate, 12-13 x 19-21 um wide; paraphyses 8 um wide at apex .... *Pachyella babingtonii*

3b Apothecia fleshy, small or large; asci bluing more toward or at apex than elsewhere; ascospores eguttulate or 1-3 guttulate, hyaline or brown 4

4a Ascospores globose... 5

4b Ascospores elliptical 6

5a Apothecia dark brown, cupulate, often up to 7 cm wide, brittle; hymenium smooth; ascospores smooth, hyaline, contents granular, irregularly biseriate at first, uniseriate later, 8-10 um wide; conidial

state *Chromelosporium*; saprophytic on burned soil, spring, widespread
.... *Plicaria endocarpoides*

5b   Apothecia blackish, discoid, smaller, up to 2 cm wide; hymenium with small warts; spores verrucose, hyaline to pale brown, 12-16 um wide; conidial stage *Chromelosporium*; saprophytic on burned soil, spring
.... *Plicaria trachycarpa*

6a   Hymenium with dark purplish to brownish purple cast, or with violaceous tints; typically on burned (heated) soil 7

6b   Hymenium without violet tints, usually yellow brown, tan, chocolate or wood brown, never with carotinoids (bright yellow or orange, red); on soil, duff, etc., occasionally on burned soil 8

7a   Hymenium purple to purple brown; exterior minutely scurfy, ascospores verrucose, hyaline, biguttulate with two polar granules, elliptical, 7-8 x 12-14 um; paraphyses with purple granules, widespread, fruiting throughout the year .... *Peziza praetervisa*

7b   Hymenium violet to reddish-violet; exterior minutely pustulate near margin and paler; ascospores smooth, hyaline, eguttulate, elliptical, 8-10 x 16-17 um; paraphyses filled with brownish granules; widespread, but not as common as *P. praeterversa*, spring or summer ....
.... *Peziza violacea*

8a   Ascospores guttulate, slightly verrucose at times 9

8b   Ascospores eguttulate, smooth to rough 13

9a   Hymenium dark reddish brown to olive brown; exterior reddish brown and scurfy; spores verrucose with warts elongating to form incomplete reticulum 10

9b   Hymenium and exterior cinnamon to pale tan; exterior smooth or pustulate; spores smooth or if rough then verrucose but without incomplete reticulum 11

10a Apothecia small, up to 18 mm wide; ascospores hyaline, with one or two unequal guttulae, verrucose, with incomplete cyanophilous reticulum, elliptical, 8-12 x 16-20 um; paraphyses unbranched, filiform; on soil, summer .... *Peziza brunneoatra*

10b Apothecia large, up to 8 cm wide, ascospores hyaline, usually with two unequal guttulae, verrucose with incomplete reticulum, elliptical, 9-12 x 17-20 um; paraphyses straight, slightly clavate; on the ground summer, fall .... *Peziza badia*

11a Spores warty or rough (when stained with cotton blue or Melzer's reagent) 12

11b Ascospores smooth; apothecia cupulate at first then repand; hymenium, umbilicate, whitish at first, later dingy-buff to brownish; ascospores ellipsoid, hyaline, biguttulate, 6-9 x 11-14 um; paraphyses slender, septate, slightly enlarged at apex; often in cellars, caves, greenhouses etc. widespread, but not often encountered ..... *Peziza domiciliana*

12a Apothecia up to 5 cm wide; whitish subiculum more or less present at base; hymenium and exterior cinnamon; ascospores hyaline, with 1-2 guttulae, elliptical, with truncate ends, delicately warty (no reticulum),

8-10 x 17-21 um; saprophytic on wood and woody conifer litter, sometimes soil, widespread in conifer woods, spring, summer . . . .
. . . . *Peziza badioconfusa*

12b Apothecia up to 10 cm wide; hymenium pale tan; exterior whitish;
ascospores hyaline, with 2 polar guttulae, minutely verrucose,
elliptical, 6-9 x 11-14 um, saprophytic on moist soil, plaster, cement
etc. . . . . *Peziza domiciliana* and *Peziza petersi*

13a Apothecia on dung or well manured soil, pale yellow-brown, sessile,
cupulate, up to 5 cm wide; margin at times incurved; exterior scurfy;
ascospores smooth, hyaline, elliptical, 10-14 x 18-24 um; spores irregularly biseriate to obliquely uniseriate, widespread, fruiting throughout
the year . . . . *Peziza vesciculosa*

13b Apothecia on soil, rotten wood, duff etc. not on dung 14

14a Apothecia large, up to 12 cm wide, sessile to short stipitate; hymenium
tan; exterior whitish, pruinose; ascospores permanently smooth;
eguttulate, hyaline, 9-11 x 15-18 um; paraphyses slender up to 7 um
wide at apex, on soil and rotting wood throughout the year . . . .
. . . . *Peziza repanda*

14b Apothecia large or small; hymenium tan to brown; spores verrucose
at maturity; paraphyses slender or inflated 15

15a Paraphyses inflated, up to 14 um wide; spores smooth, to verrucose
(at maturity), eguttulae, hyaline, elliptical, 9-11 x 14-19 um; on soil or
rotting wood, throughout the year . . . . *P. varia*

15b Paraphyses not inflated, up to 8 um wide or less 16

16a Exterior coarsely pustulate; apothecia small to large, up to 6 cm wide,
shallow cupulate, sessile; hymenium pale brown; exterior whitish;
ascospores smooth to minutely verrucose, eguttulate, elliptical, 7-12 x
13-18 um; paraphyses filiform to clavate; ectal excipulum thin, of
globoid to polyhedral cells; on soil and charcoal throughout the year,
not common . . . . *Peziza echinospora*

16b Exterior finely pustulate; apothecia large, up to 8 cm wide, cupulate,
sessile; hymenium medium brown; exterior whitish to tan; ascospores
smooth to finely verrucose, eguttulate, 9-10 x 15-20 um; paraphyses
clavate; ectal excipulum not distinguished; on soil or duff throughout
conifer forests, spring . . . . *Peziza sylvestris*

BAY COLORED FAIRY CUP
Peziza badia Pers.

FRUITING BODIES: Scattered to cespitose; cupulate, margin wavy, sessile, up to 8 cm wide; *hymenium* dark brown, smooth; *exterior*, reddish brown, scurfy; *mycelium* white.

MICROCHARACTERS: *Spores* 15-20 x 7-10 um, elliptical, rough with irregular longitudinal broken ridges forming an incomplete reticulum, typically biguttulate and with one droplet larger than the other. *Asci* 270-350 x 15 um, J+. *Paraphyses* clavate, septate, slightly roughened and 5 um wide at apex. *Medullary excipulum* of interwoven hyphae up to 10 um wide and scattered globose cells up to 35 um wide. *Ectal excipulum* of gelatinous tissues with outer layers of globose, oblong or polyhedral cells.

HABITAT-DISTRIBUTION: Widespread in coniferous forests of temperate zone, on soil or humus rarely on wood, July, August, September, and October, common.

DESCRIPTIONS: refs. 6, 10, 37.

REMARKS: Edibility unknown. Often mistaken for *P. badioconfusa* which has warty spores, fruits vernally and occurs on woody coniferous debris. A comparison of the two species has been made by Elliott and Kaufert (10).

CONFUSING FAIRY CUP
Peziza badioconfusa Korf

FRUITING BODIES: Scattered to clustered; cupulate to otidioid or flat, sessile on a sparce subiculum, up to 5 cm wide; *hymenium* brownish orange or darker, smooth; *exterior* colored like hymenium, smooth to slightly scurfy and dull; *mycelium* white, forming a subiculum

MICROCHARACTERS: *Spores* 17-21 x 8-10 um, elliptical with more or less truncate ends, delicately warty, with one or two oil drops. Asci 210-235 x 13-15 um wide, J+. *Paraphyses* clavate, up to 7 um wide at apex. *Excipulum* similar to *P. badia*.

HABITAT-DISTRIBUTION: On rotting wood, or conifer litter, rarely on soil in conifer forests throughout U.S. and Canada, late spring and summer.

DESCRIPTIONS: refs. 6, 10, 31, 37.

REMARKS: This species has been or is easily confused with *P. badia*. The vernal fruiting habit, the somewhat truncate ascospores and the fine warts rather than an incomplete reticulum distinguish it. Edibility not tested.

BLACKISH BROWN FAIRY CUP
Peziza brunneoatra Desmazieres

FRUITING BODIES: Scattered to clustered; cupulate at first, flattening at maturity, sessile, small, up to 18 mm wide; *hymenium* blackish brown; *exterior* dark reddish brown, paler below, smooth or delicately scurfy at the margin.

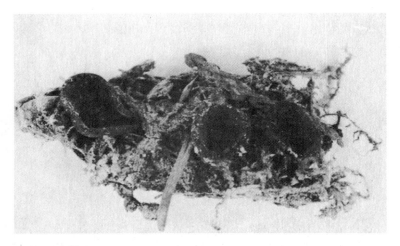

MICROCHARACTERS: *Spores* 16-20 x 8-12 um, elliptical, hyaline, warty, the large warts cyanophilic and forming an incomplete reticulum, with one or two oil drops. *Asci* 260-300 x 12-16 um, J+. *Paraphyses* filiform to clavate, occasionally septate, contents oily brown, up to 6 um wide at apex. *Medullary excipulum* of globose to polyhedral cells up to 60 um wide and branched pale brown septate hyphae; *Ectal excipulum* of smaller polyhedral, rectangular or cuboidal cells.
HABITAT-DISTRIBUTION: Widespread on soil throughout North America, June and July, not common.
DESCRIPTIONS: refs, 6, 31, 37.
REMARKS: This rather small species is unattractive for the table. Edibility not tested.

PETER'S PEZIZA
Peziza petersii Berk. & Curt.

FRUITING BODIES: Scattered to gregarious; up to 4 cm wide, cupulate to repand; *hymenium* unevenly colored, light grayish reddish brown to medium brown near margin, confoluted or wrinkled; *exterior* pale grayish lilac, smooth; *stipe* short or absent.

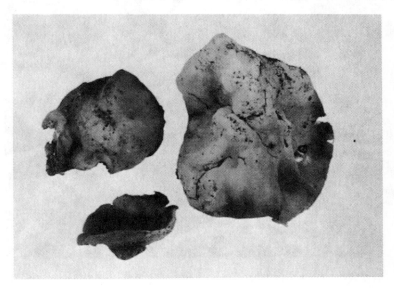

MICROCHARACTERS: *Spores* 10-12 x 5-6 um, elliptical, with delicate warts, biguttulate. *Asci* 190-200 x 10 um, J+. *Paraphyses* slightly clavate, often curved, unbranched, with irregular brownish globules, up to 7 um wide at apex.
HABITAT-DISTRIBUTION: Under conifers in spruce fir zone, at times on burned ground, summer and fall, not common.
DESCRIPTION: ref. 6.
REMARKS: Edibility untested. Distinguished from similar Pezizas of like habitat by the finely warted and biguttulate ascospores.

PURPLE FAIRY CUP
Peziza praetervisa Bres.

FRUITING BODIES: Scattered to gregarious; cupulate, flattening, sessile, up to 3 cm wide; *hymenium* purple to purple brown; *exterior* pale purple, finely scurfy.

MICROCHARACTERS: *Spores* 12-14 x 7-8 um, elliptical, finely warted, with two polar granules. *Asci* 250-300 x 10-12 um, J+. Paraphyses clavate, branched, filled with violet granules, tips bent and up to 7 um wide. *Medullary excipulum* of elongate and globose cells up to 60 um wide. *Ectal excipulum* of globose and polyhedral cells, occasionally some in chains.

HABITAT-DISTRIBUTION: Widespread throughout the United States and Canada on burned soil, most often found on remains of old campfires.

DESCRIPTIONS: refs. 6, 31, 37.

REMARKS: This species has been confused with the violet fairy cup *P. violacea*. It is separated from the latter by its dark purple hymenium and biguttulate warty ascospores. Edibility untested. Another species with purple brown hymenium is *Pachyella babingtonii*. It is distinguished by its small size, translucent appearance and an ascus which blues throughout.

SPREADING PEZIZA
Peziza repanda Pers. ex Fr.

FRUITING BODIES: Scattered to clustered; cupulate then spreading flat, sessile or short stipitate, large, up to 12 cm wide; *hymenium* pale brown, even or convoluted; *exterior* whitish, pruinose.

MICROCHARACTERS: *Spores* 15-18 x 19-21 um, elliptical, smooth, without oil drops, *Asci* 250-300 x 12-15 um, J+. *Paraphyses* slightly clavate to filiform, unbranched up to 7 um wide at apex. *Medullary excipulum* three layered, the central layer of interwoven hyphae somewhat parallel to hymenium and the two other layers of large globose or polyhedral cells and randomly oriented hyphae. *Ectal excipulum* of textura angularis and external heaps of globose to ellipsoid cells.
HABITAT-DISTRIBUTION: Widespread throughout the conifer forests and in gardens, spring, summer and fall, common.
DESCRIPTIONS: refs. 6, 31, 37.
REMARKS: It is very close to *P. varia* which has rough spores. *Discina* species could be confused with this *Peziza* if field characters alone are used. In *Discina* the spores are apiculate, and fruiting occurs in the early spring. Edibility not tested.

SYLVAN PEZIZA

Peziza sylvestris (Boud.) Sacc. and Trott.

FRUITING BODIES: Gregarious; cupulate, margin wavy, sessile, large, up to 8 cm wide; *hymenium* medium brown, convoluted; *exterior* tan to whitish, finely pustulate, flesh thick.

MICROCHARACTERS: *Spores* 15-20 x 9-10 um, elliptical, smooth or finely verrucose, eguttulate. *Asci* 175-250 x 12-14 um, J+. *Paraphyses* clavate, the tips up to 7 um wide. *Medullary excipulum* of globose cells, and densely interwoven swollen hyphae. *Ectal excipulum* not well differentiated.

HABITAT-DISTRIBUTION: Widespread in coniferous woods, on litter and soil, spring.

DESCRIPTIONS: refs. 31, 37.

REMARKS: *Peziza echinospora* is very similar but is distinguished by the paler hymenium and coarse pustules on the margin. According to LeGal this fungus is the same as *Peziza arvernensis*. Edibility untested.

VARYING PEZIZA
Peziza varia Hedw. ex Fr.

FRUITING BODIES: Single to gregarious or crowded; cupulate to discoid, sessile to short stipitate, margin often crenulate, up to 6 cm wide; *hymenium* gray to dark grayish brown, even; *exterior* whitish, pruinose, and typically pustulate near margin.

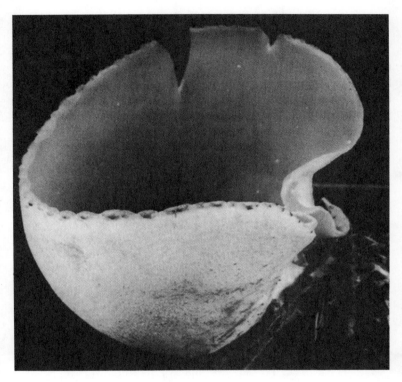

MICROCHARACTERS: *Spores* 14-19 x 9-11 um, elliptical, smooth or minutely verrucose, eguttulate. *Asci* 225-300 x 12-16 um, J+. *Paraphyses* moniliform, apex inflated up to 14 um wide. *Hypothecium* 40-60 um thick, of textura angularis. *Medullary excipulum* of three layers, the middle layer up to 100 um wide, of interwoven hypha approximately parallel to hymenium, the other two layers of large globose to polyhedral cells and randomly oriented hyphae.
HABITAT-DISTRIBUTION: Fruits year round on rotting wood, duff, or soil often in residential basements, widespread throughout

the United States.

DESCRIPTIONS: refs. 6, 31.

REMARKS: This species resembles *P. repanda* which has smooth spores and paraphyses which are not moniliform or greatly inflated. Edibility untested.

BLISTERED PEZIZA
Peziza vesciculosa Bull.

FRUITING BODIES: Scattered to gregarious; cupulate with incurved margin, sessile, up to 5 cm wide; *hymenium* light yellow brown, becoming detached at times and forming a blister; *exterior* pale tan, scurfy.

MICROCHARACTERS: *Spores* 18-24 x 10-14 um, elliptical, smooth, eguttulate. *Asci* 230-330 x 14-18 um, J+, cylindric with abruptly tapered base. *Paraphyses* filiform, septate, scarcely enlarged at apex, up to 4 um wide. *Hypothecium* textura angularis. *Medullary excipulum* of narrow hyphae, globose to ellipsoid cells and scattered lactifers. *Ectal excipulum* poorly differentiated, forming clusters of globoid cells.

HABITAT-DISTRIBUTION: Common, on dung or manured soil, widespread throughout the United States, June and July but fruiting year round as conditions permit.

DESCRIPTIONS: refs, 6, 31, 37.

REMARKS: The fruiting bodies on dung tend to be larger than those on soil. Edibility not tested. *Peziza domiciliana* is similar but is distinguished by the smooth exterior, smaller spores and habitat.

VIOLET FAIRY CUP
Peziza violacea Pers.

FRUITING BODIES: Scattered to gregarious; cupulate then expanding, sessile, up to 3 cm wide; *hymenium* pale violet to reddish violet, flesh thin, pale purple; *exterior* paler, grayish, delicately pruinose near margin.

MICROCHARACTERS: *Spores* 16-17 x 8-10 um, elliptical, smooth, eguttulate. *Asci* 300-340 x 10-11 um, J+. *Paraphyses* clavate, filled with brown granules, tips bent, up to 7 um wide. *Hypothecium* textura intricata. *Medullary excipulum* of globose cells oriented more toward exterior and intermixed elongate cells. *Ectal excipulum* of polyhedral to globose cells up to 15 um wide.

HABITAT-DISTRIBUTION: Widespread on burned soil, infrequently encountered, spring or summer.

DESCRIPTIONS: refs. 6, 31, 37.

REMARKS: Resembles *P. praetervisa* another violet species that grows on burned soil but which has rough spores with two polar oil drops and a dark purple to purple brown hymenium. Edibility not tested.

SMOOTH FAIRY CUP
Plicaria endocarpoides (Berk.) Rifai

FRUITING BODIES: Scattered to gregarious; cupulate then spreading flat, sessile, brittle, up to 7 cm wide; *hymenium* dark brown; *exterior* paler, smooth.

MICROCHARACTERS: *Spores* 8-10 um wide, globose, smooth, with granular contents. *Asci* 190-200 x 9.5-11 um wide, J+. *Paraphyses* filiform, slightly expanded at apex, up to 4 um wide and covered by an amorphous yellow-brown encrustation. *Medullary excipulum* of textura globulosa and filamentous cells, up to 2 um thick. *Ectal excipulum* of textura globulosa, the cells smaller than in medullary excipulum.

HABITAT-DISTRIBUTION: Widespread, on burned soil, spring.

DESCRIPTIONS: refs. 6, 31, 37.

REMARKS: Edibility unknown. It is separated from *P. trachycarpa* which it resembles by the smooth spores, more persistently cupulate, larger apothecia and the dark brown rather than blackish hymenium color.

ROUGH FAIRY CUP
Plicaria trachycarpa (Currey) Boud.

FRUITING BODIES: Scattered to gregarious; discoid, sessile up to 2 cm wide; *hymenium* blackish, rough with small warts; *exterior* dark brown, scurfy.

MICROCHARACTERS: *Spores* 12-16 um wide, globose, finely warted, hyaline to pale brown. *Asci* 320-350 x 17-20 um wide, J+. *Paraphyses* clavate, the tips up to 8.5 um wide, encrusted with a yellowish amorphous material. *Medullary excipulum* a mix of hyphal cells and globose to polyhedral cells. *Ectal excipulum* of smaller globose to polyhedral cells.

HABITAT-DISTRIBUTION: Widespread on burned soil, spring and fall, not common.

DESCRIPTIONS: refs. 6, 31, 37.

REMARKS: Differs from *P. endocarpoides* another species with globose spores by the verrucose ascospores. Edibility untested.

VIOLET STAR CUP
Sarcosphaera crassa (Santi) Pouzar

FRUITING BODIES: Single to gregarious; at first a white, globose, hollow and subterranean ball, becoming erumpent and cupulate with a star-shaped split margin at maturity, large, up to 15 cm wide, sessile or with short stalk; *hymenium* violet, smooth; *exterior* whitish to cream colored especially at base, finely scurfy.

MICROCHARACTERS: *Spores* 15-22 x 7-9 um, ellipsoid but with somewhat truncate ends, smooth or nearly so, with 1-3 oil drops. *Asci* 350-375 x 15-17 um wide, J+ tip. *Paraphyses* narrow below, with expanded or knobbed apex up to 10 um wide, occasionally curved, septate and slightly constricted at septa, with purple-brown granular contents. *Hypothecium* of textura angularis about 50 um wide. *Medullary excipulum* up to 3 mm wide, of densely interwoven stout hyphae. *Ectal excipulum* of textura intricata.

HABITAT-DISTRIBUTION: Widespread, in conifer and hardwood forests on or in the soil, duff or litter, common in Pacific Northwest, spring and early summer.

DESCRIPTIONS: refs. 6, 31.

REMARKS: The early, subterranean habit and star-shaped margin with internal violet colors are distinctive. Not edible and can be poisonous to some people (23). *Peziza ammophila* is similar but occurs in sandy habitats and has a dark brown hymenium.

6. Helvellaceae

This family includes the poisonous false morels as well as a number of excellent edible Discomycetes. Unfortunately the family is defined primarily on microscopic characters so that very few field characters are shared by the group. The fruiting body shape varies from cupulate or discoid to the typical well-developed stalked apothecia with cupulate, gyrose, saddle-shaped or miter-like caps.

Of the short stalked or sessile, cupulate forms, both *Wynnella* and *Rhizina* are rare, but *Discina* species are common and these will be difficult to distinguish on a field basis from other cup fungi such as the Pezizaceae. The early spring fruiting habit of *Discina* species is perhaps the most useful distinguishing feature. The stalked species are different enough to be readily identified.

Many species in this family have been found to produce the toxin gyromitrin or monomethylhydrazine. Therefore, none of the species in this family should be eaten raw and before eating they should be parboiled and the water discarded. Also be sure of your identification and do not experiment because the distribution of the toxin in the species of this family has not yet been determined.

Diagnostic Description
Apothecia usually pileate, large and saddle-shaped or gyrose, sometimes cupulate or otidioid, black, gray or brown, without carotinoids; asci not blue in iodine; ascospores smooth, warted or apiculate, tetranucleate, hyaline or brown.

LIST OF SPECIES
1. *Discina apiculatula* McKnight
2. *Discina leucoxantha* Bres.
3. *Discina macrospora* Bubak
4. *Discina perlata* (Fr.) Fr.
5. *Gyromitra ambigua* (Karst.) Harmaja
6. *Gyromitra californica* (Phillips) Raitviir
7. *Gyromitra esculenta* Fr.
8. *Gyromitra gigas* (Kromb.) Quel.
9. *Gyromitra infula* (Schaeff. ex Fr.) Quel.
10. *Helvella acetabulum* (L. ex St. Amans) Quel.
11. *Helvella albipes* Fuckel
12. *Helvella corium* (Weberb.) Massee
13. *Helvella crispa* Scop. ex Fr.
14. *Helvella elastica* Bull. ex St. Amans
15. *Helvella griesoalba* Webber

16. *Helvella lacunosa* Afz. ex Fr.
17. *Helvella leucomelaena* (Pers.) Nannf.
18. *Helvella macropus* (Pers. ex Fr.) Karst.
19. *Helvella queletii* Bres.
20. *Helvella villosa* (Hedw. ex Kuntze) Dissing & Nannfeldt
21. *Paxina recurvum* Snyder
22. *Peziza melaleucoides* Seaver
23. *Rhizina undulata* Fr.
24. *Wynnella silvicola* (Beck. ex Sacc.) Nannf.

KEY

1a Apothecia discoid, cupulate repand, otidioid, sessile, or if short stipitate pileus cupulate 2

1b Apothecia stipitate-pileate, pileus saddle-shaped, mitrate, convoluted or gyrose, or long stipitate and cupulate 8

2a Apothecia elongate ear-shaped (otidioid), up to 8 cm high, yellowish at first, blackish brown in age, pale brown at base, asci not bluing; ascospores 20-25 x 12-16 um; smooth, broadly ellipsoid with one central oil drop; paraphyses with pale brown granules, clavate at apex, rare . . . . *Wynnella silvicola*

2b Apothecia not otidioid 3

3a Apothecia sessile, undulate, crust-like, attached to substrate over the entire lower side by numerous yellow rhizoids, up to 6 cm wide; hymenium, dark purple-brown with yellow, irregular margin; asci 450-500 x 11-14 um; ascospores apiculate, fusoid, 30-40 x 8-11 um, smooth at first, short verrucose at maturity; paraphyses brown encrusted; tubular setae present; on burned soil, at times parasitic on conifer seedlings . . . . *Rhizina undulata*

3b Apothecia not crust-like and with rhizoids on entire lower surface 4

4a Underside of apothecium with prominent branching ribs (like a cabbage leaf) 19

4b Apothecia sessile or short stipitate; expanded cupulate portion without branched ribs on lower side 5

5a Apothecium brownish-black (smoky) whitish below, cupulate, up to 5 cm wide, short stipitate; stipe whitish, with roundish ribs which do not extend up onto the cup; ascospores elliptical, hyaline, smooth, with one oil drop, 20-24 x 10-13 um; paraphyses clavate, sometimes branched, up to 7 um wide at tip; on soil, common . . . . *Helvella leucomelaena*

5b Apothecia not blackish, some shade of light brown to yellow brown, or reddish brown, discoid, repand or cupulate 6

6a Apothecia sessile, cupulate to repand; hymenium yellow or yellowish brown; ascospores apiculate, elliptical, with one large central oil drop and 2 small polar ones, smooth at first, warty at maturity, 14-16 x 31-35

um; apiculi blunt, somewhat cupulate; on soil, not common . . . .
. . . . *Discina leucoxantha*

6b   Hymenium brown to yellow brown at maturity; apothecia short stipitate, cupulate to repand 7

7a   Hymenium moderate or strong brown, wrinkled; ascospores 11-16 x 25-35 um, elliptic to broadly fusoid, smooth then delicately verrucose with usually one central oil drop, apiculate; apiculus short 1-3.5 um; developing tardily; on soil, duff, or rotting wood, common . . . .
. . . . *Discina perlata* and *Discina macrospora*

7b   Hymenium yellow brown, convoluted; ascospores 13-18 x 28-33 um; smooth, becoming minutely verrucose-reticulate, apiculate; apiculi short truncate to obtuse or broadly rounded; on soil in spruce-fir forests . . . . *Discina apiculatula*

8a   Fruiting body gyrose (brain-like) 9

8b   Fruiting body saddle-shaped, cupulate, mitrate or umbricauliform
. . . . 10

9a   Stipe very short, thick, nearly equaling pileus in width, whitish, lacunose; pileus up to 10 cm wide, medium brown to yellowish brown; ascospores ellipsoid-fusoid, apiculate at maturity, smooth to minutely verrucose, with one central and two polar oil drops, 11-15 x 24-45 um; paraphyses up to 10 um wide at apex, clavate; on soil and litter, often near snowbanks. . . . . *Gyromitra gigas*

9b   Stipe prominent, slender, pale reddish tan, hollow, not lacunose; pileus dark reddish brown, up to 8 cm wide; ascospores smooth, without apiculus, elliptical, biguttulate, 20-26 x 10-13 um; paraphyses 3-4 septate, clavate, up to 8 um wide at tip, red-brown in dilute KOH; on the ground, rarely on decaying wood, common . . . .
. . . . *Gyromitra esculenta*

10a   Stipe ribbed 11

10b   Stipe terete, or compressed, not ribbed except for few basal clefts..14

11a   Pileus umbricauliform; spores smooth at first, apiculate only at full maturity; pileus light to olive brown, somewhat gyrose; stipe short, broad, often tinged pink especially near base; ribs prominent, extending to margin of pileus; ascospores ellipsoid-fusoid with two polar oil drops, smooth, 7-10 x 13-19 um; paraphyses clavate up to 10 um wide at apex; on soil or rotting wood, common . . . . . . *Gyromitra californica*

11b   Pileus saddle-shaped, or mitral; spores non-apiculate 12

12a   Pileus and stipe pure white to pale ivory, saddle-shaped to convolute with reflexed margin; stipe strongly ribbed attached at apex only, lacunose; ascospores hyaline, smooth, elliptical with one central and several polar guttulae, 11-14 x 18-24 um; paraphyses up to 10 um wide at apex; on soil and duff . . . . *Helvella crispa*

12b   Pileus colored gray brown, to black, at times stipe also colored . . 13

13a   Pileus irregularly cupulate, up to 8 cm broad, brown; stipe short, tan; ribs at times branching, but not extending up on to cupulate portion; ascospores smooth, ellipsoid, with single central droplet and several small polar ones, 20-24 x 11-14 um; paraphyses up to 8 um wide at apex;

on soil or duff, widespread, not common .... *Helvella queletii*
13b Pileus irregularly convex or mitrate, dark gray to nearly black, attached to stipe at or near the margin at several points; stipe long, pale to gray or blackish, strongly ribbed, lacunose; ascospores smooth, elliptical with slightly truncate ends, with one large central oil drop, 12-14 x 17-21 um; paraphyses occasionally branched, clavate, up to 9 um wide at apex, sometimes curved; on rotting wood or soil, common ....
.... *Helvella lacunosa*
14a Pileus cupulate 15
14b Pileus convex, saddle-shaped 17
15a Hymenium smoky gray to black 16
15b Hymenium gray brown to buffy brown; apothecia up to 5 cm broad, permanently cup-shaped, fibrillose scaly, particularly near margin; stipe solid, tapering toward apex, concolorous; ascospores fusiform, verrucose when young, with one large central oil drop and single smaller polar ones, 20-25 x 10-12 um; paraphyses clavate up to 10 um wide at apex; on soil, duff.................... *Helvella macropus*
16a Apothecia cupulate, margin spreading, crenate or lobed, black; exterior furfuraceous; stipe even, black above, gray at the base, terete with rounded ribs at base; ascospores smooth, ellipsoid, with one large central oil drop, 17-21 x 10-12 um; pigment on hyphal walls continuous; on the ground .... *Helvella corium*
16b Apothecia cupulate at first, nearly plane or repand in age, gray-black, margin entire or lacerate in age, exterior delicately velvety hairy, concolorous; stipe terete or compressed, tapering toward apex, concolorous but paler at base, delicately velvety hairy especially at apex; ascospores ellipsoid, smooth or verrucose, with prominent central guttule, 16-19 x 13-14 um; pigment on excipular hyphae in uneven patches; on the ground .... *Helvella villosa*
17a Hymenium reddish brown; paraphyses and medullary excipulum staining reddish brown in dilute KOH; apothecia 6-20 cm high, saddle-shaped or mitral attached to stipe at several points along the margin; stipe hollow, pinkish tan, minutely tomentose; ascospores smooth, biguttulate, 7-9 x 17-22 um; paraphyses with bulbous tips up to 14 um wide, sometimes staining red brown in KOH; on rotting wood or duff, common, fall .... *Gyromitra infula* and *Gyromitra ambigua*
17b Hymenium gray or tan, tissue of excipulum or hymenium not deep reddish brown in KOH 18
18a Pileus turned down on three or four sides and thus having an angular appearance, up to 10 cm wide, yellow brown; reverse pale; stipe smooth, 1-2 cm in dia. up to 4 cm high, lacunose internally; ascospores elliptical to oval, hyaline, verrucose, with 1-3 oil drops (usually two) 7-10 x 12-14 um; paraphyses stout up to 10 um wide at apex; on soil .... *Paxina recurvum* and *Peziza melaleucoides*
18b Pileus saddle-shaped to mitral, gray brown; stipe solid, smooth, 1 cm or less in diameter, up to 6-8 cm long, tapering toward apex, cream

colored; ascospores elliptical, smooth or at times verrucose, with one large central oil drop 10-14 x 18-22 um; paraphyses clavate, occasionally branched; on soil or duff ....

.... *Helvella elastica* and *Helvella albipes*

19a Apothecium up to 6 cm wide, hymenium brown, exterior cream; ascospores elliptical, hyaline, uniguttulate, 18-20 x 12-14 um; paraphyses enlarged up to 10 um wide near apex; ectal excipulum cells with diffuse brown cytoplasmic pigment, in duff, rare .... *H. acetabulum*

19b Apothecium up to 3-4 cm wide, hymenium grayish to white; ascospores 15-18 x 10-12 um, oblong or broadly ellipsoid, uniguttulate, hyaline; paraphyses narrowly clavate up to 6 um wide at apex; ectal excipulum cells without cytoplasmic pigments, but light brown pigment often found in projecting elements; in duff, rare .... *H. griseoalba*

Foray Activity: Dr. Alexander Smith examines a rare mushroom collected during a foray in Idaho while Kit Scates (center), Dr. Hard (right), and others look on.

PIG'S EAR
Discina apiculatula McKnight

FRUITING BODIES: Scattered to clustered; at first cupulate, then expanding to repand, margin inrolled at first, sessile or short stipitate, up to 12 cm wide; *hymenium* moderate yellowish brown, wrinkled; *exterior* light brown, glabrous or nearly so.

MICROCHARACTERS: *Spores* 28-33 x 13-18 um, ellipsoid, smooth, becoming verrucose-reticulate, apiculus short, rounded, with one large central and several smaller oil globules. *Asci* 270-300 x 13-18 um, J-. *Paraphyses* clavate, septate, often lobed at apex, up to 7 um wide at apex. *Medullary excipulum* textura intricata, oleiferous hyphae present. *Ectal excipulum* textura intricata, more compact toward the interior.

HABITAT-DISTRIBUTION: On soil or duff, under spruce-fir, spring.

DESCRIPTION: ref. 27.

REMARKS: Scarcely distinguishable from *Discina perlata* on field characters. Edibility untested but probably has been eaten by those collecting and eating *Discina perlata*.

YELLOWISH DISCINA
Discina leucoxantha Bres.

FRUITING BODIES: Single to gregarious; cupulate at first then repand, substipitate to sessile, up to 10 cm wide; *hymenium* pale yellowish brown, surface with shallow irregularities (undulate) in age; *exterior* yellowish, glabrous.

MICROCHARACTERS: *Spores* 31-35 x 14-16 um, narrowly ellipsoid, smooth at first, then verrucose. The warts forming an interrupted reticulum; with one large central and two smaller polar drops which often merge with the central one; apiculus becoming depressed at apex. *Asci* 390-475 x 20-25 um, attenuated at base, J-. *Paraphyses* clavate, with ochraceous granular contents, up to 10 um wide at apex. *Medullary excipulum* of loose textura intricata. *Ectal excipulum* more compact, textura intricata gradually merging with the medullary excipulum.

HABITAT-DISTRIBUTION: Rare, on conifer duff, collected only once in Idaho from Brundage Mtn, spring.

DESCRIPTIONS: refs. 27, 31.

REMARKS: Edibility unknown. The yellow hymenium separates this *Discina* from all other species which are typically brown. Microscopically the invaginated apiculus is diagnostic.

PIG'S EAR
Discina perlata Fr.

FRUITING BODIES: Scattered to clustered; cupulate, expanding, often umbilicate and wrinkled or convoluted, sessile or short stipitate, up to 6 cm wide; *hymenium* at first strong yellow brown especially if encountered growing under snow, later dark yellow brown, wrinkled; *exterior* pale, smooth.

MICROCHARACTERS: *Spores* 25-35 x 11-15 um, ellipsoid to broadly fusoid, smooth at first, later delicately verrucose, with one large central and several polar oil drops, apiculi tardily developed, rounded at first, becoming fully pointed only at full maturity, up to 3 um long. *Asci* 350-400 x 17-21 um, cylindric, J-. *Paraphyses* cylindric, stout, clavate apex up to 10 um wide, contents of brown granules. *Medullary excipulum* loosely woven textura intricata, the hyphae up to 24 um wide, *Ectal excipulum* a narrow band of loose textura intricata.

HABITAT-DISTRIBUTION: On soil, duff or rotting wood under conifers, early spring, common, widespread.

DESCRIPTIONS: refs. 6, 27, 31, 37.

REMARKS: Edible but usually of too little substance to interest most collectors especially as the snow mushroom is available at the same time. *Discina apiculatula* and *Discina macrospora* are indistinguishable from *D. perlata* on field characters. *D. macrospora* has large acutely apiculate spores 27-34 um long, whereas *D. apiculatula* has rounded apiculate spores. *D. perlata* produces immature pale colored fruiting bodies under the snow and several weeks are required, after the snow recedes, for the development of the dark yellow brown hymenium. The spores mature slowly, and pass through a rounded apiculus phase so that immature material could be identified as *D. apiculatula*.

CALIFORNIA ELFIN SADDLE
Gyromitra californica (Phillips) Raitviir

FRUITING BODIES: Solitary to gregarious; up to 12 cm wide; *pileus* olive brown, undulate spreading, umbrella-shaped, margin inrolled, underside white to cream, smooth; *stipe* whitish with a flush of pink, fluted, the ribs extending to margin of pileus, up to 3 cm wide and 9 cm long.

MICROCHARACTERS: *Spores* 13-19 x 7-10 um, ellipsoid-fusoid, smooth, with two small polar guttulate. *Asci* 225-260 x 13-15 um, J-. *Paraphyses* clavate, up to 10 um wide at apex. *Medullary excipulum*

of textura globulosa and filamentous hyphae.

HABITAT-DISTRIBUTION: Common, and widespread in Western U.S. especially the Pacific Northwest and Rocky Mountains, late spring and summer on soil and duff in open woods, skid roads and margins of forests occasionally on rotting coniferous wood.

DESCRIPTIONS: refs. 31, 37.

REMARKS: The pinkish tinge on the stipe is variable. Occasionally the entire stipe is rosy, but most often only the very base is pink. I have two records of consumption of Idaho collections of this fungus with no ill effects. However, consumption of *Gyromitra* species is not recommended because of the possibility of gyromitrin poisoning. Miller (29) lists it as poisonous.

BRAIN MUSHROOM OR CALF'S BRAIN
Gyromitra esculenta Fr.

FRUITING BODIES: Scattered to gregarious; *pileus* deep reddish brown to dark brown, convoluted, brain-like, overall somewhat globose, margin attached in several places up to 8 cm wide; *stipe* tan, or flesh colored, short and stout, hollow, up to 5 cm long and 2.5 cm wide.

MICROCHARACTERS: *Spores* 20-26 x 10-13 um, elliptical, smooth but occasionally with perispore at apices, biguttulate. *Asci* 300-340 x 16-18 um, J-. *Paraphyses* clavate, stout, 3-4 septate, reddish in KOH, apices up to 8 um wide. *Medullary excipulum* loose interwoven textura intricata. *Ectal excipulum* hardly differentiated.

HABITAT-DISTRIBUTION: Common and widespread throughout U.S. and Pacific Northwest on the ground under conifers and hardwoods.

DESCRIPTIONS: refs. 6, 31, 37, 39.

REMARKS: Edibility debatable. Most mushroom books list it as poisonous, mainly because people from Europe and Eastern U.S. have died from it. Yet this fungus has been eaten by many people in the Western U.S. for years with no ill effects. It is possible the western race contains less of the toxin gyromitrin than the eastern ones. This toxin evaporates at 80° C (below the boiling point of water) and may be removed on cooking. Hence caution is indicated, and certainly do not eat this fungus unless adequately cooked.

In the field: The author collecting a "find" of the Walnut near Harvard in Northern Idaho. Note equipment and collecting technique.

WALNUT OR SNOW-MUSHROOM
Gyromitra gigas (Kromb.) Quelet

FRUITING BODIES: Single to gregarious; *pileus* strong yellow brown, convoluted, gyrose, attached to stipe near margin, up to 10 cm wide; *stipe* whitish, short, thick, almost as wide as cap, channeled within, 3-5 cm wide x 2-4 cm long.

MICROCHARACTERS: *Spores* 24-45 x 11-14 um, ellipsoid-fusoid, smooth to minutely verrucose, with rounded apiculi at maturity, with one large and several small polar oil drops. *Asci* 350-600 x 17-19 um, J-. *Paraphyses* clavate, stout, branched, septate with yellowish granular contents in KOH, the tips up to 10 um broad. *Medullary excipulum* of loosely interwoven textura intricata. *Ectal excipulum* hardly distinguished.

HABITAT-DISTRIBUTION: Widespread in western U.S. in early spring shortly after the snows recede or near melting snowbanks throughout the spring and summer on soil, duff or rotting wood.

DESCRIPTIONS: refs. 37, 40, 41.

REMARKS: Edible, delicious and popular with collectors throughout the Rocky Mountains. It is closely related to the Pig's Ear *Discina perlata* which can at times produce helvelloid forms which are difficult to distinguish from *G. gigas*.

HOODED GYROMITRA
Gyromitra infula (Schaeff. ex Fr.) Quel.

FRUITING BODIES: Single to scattered; *pileus* strong brown, saddle-shaped with two or three folded sections attached to the stipe at several points, up to 10 cm wide; *stipe* pale yellowish pink, minutely tomentose, hollow, up to 6 cm long.

MICROCHARACTERS: *Spores* 17-22 x 7-9 um, elliptical, smooth, biguttulate. *Asci* 300-340 x 10-12 um, J-. *Paraphyses* stout, with bulbous tips, 3-4 septate, red brown in KOH, the tips up to 14 um wide. *Medullary excipulum* of loosely interwoven textura intricata. *Ectal excipulum* hardly differentiated.

HABITAT-DISTRIBUTION: Widespread, late summer and fall on decaying wood, sawdust, duff or soil, common.

DESCRIPTIONS: refs. 6, 31, 37.

REMARKS: Poisonous, contains gyromitrin. The fall fruiting habit and paler, less gyrose cap distinguish this species from *G. esculenta*. Another fall, poisonous species indistinguishable on a field basis is *G. ambigua*. It differs in having slightly larger spores with bluntly projecting perispore, and the apices are somewhat narrower than in *G. infula*.

CABBAGE LEAF HELVELLA
Helvella acetabulum (L. ex St. Amans) Quel.

FRUITING BODIES: Scattered to gregarious; cupulate to nearly plane at maturity, the ribs of the short stipe branching and extending almost to the margin, sessile or stipitate, up to 8 cm wide; *hymenium* dark brown; smooth or slightly undulate; *exterior* concolorous with hymenium, paler near the pale cream colored stipe or base.

MICROCHARACTERS: *Spores* 18-20 x 12-14 um, broadly ellipsoid, smooth, with a single central oil drop. *Asci* 350-400 x 15-20 um, J-. *Paraphyses* clavate, pale brown, the apex up to 10 um wide. *Medullary excipulum* of textura intricata, up to 600 um thick. *Ectal excipulum* of textura angulata, outer cells diffusely brown.

HABITAT-DISTRIBUTION: Widespread on the ground in woods, spring and summer in the Pacific Northwest and Rocky Mountains, rarely encountered.

DESCRIPTIONS: refs. 6, 43.

REMARKS: Edibility unknown and not recommended since related forms contain gyromitrin. The extensive branched ribbing on the lower side distinguishes this species from all other *Helvella* species except *H. griseoalba*. *H. queletii* is similar in habit, but the ribbing in that species does not extend up to near the margin as it does in *H. acetabulum*. *H. griseoalba* differs in having a gray hymenium but is otherwise very similar.

CRISPED HELVELLA
Helvella crispa Scop. ex Fr.

FRUITING BODIES: Scattered; up to 4 cm high and 5 cm wide; *pileus* white to pale cream, saddle-shaped or lobed, attached to stipe at apex only; margin inrolled at first, flaring in age; sterile side pruinose; *stipe* white to cream, deeply fluted, lacunose internally, up to 10 cm long and 3 cm wide.

MICROCHARACTERS: *Spores* 18-24 x 11-14 um, elliptical, smooth, with one large central oil drop and several small polar ones. *Asci* 275-340 x 14-18 um, J-. *Paraphyses* narrowly clavate, branched or unbranched up to 10 um broad at apex. *Medullary excipulum* of textura intricata, the hyphae 3-4 um wide. *Ectal excipulum* of textura globulosa with fascicles or short chains of hyaline cells.

HABITAT-DISTRIBUTION: On soil, duff and moss banks in coniferous woods, common throughout the U.S. in the late summer through fall.

DESCRIPTIONS: refs. 6, 16, 31, 43.

REMARKS: Edible but not recommended, and never to be eaten raw. The possibility of gyromitrin/monomethylhydrazine poisoning is likely if any *Helvella* is misidentified. This species superficially resembles pale colored forms of *H. lacunosa* but differs in that the pileus is attached at apex of stipe only.

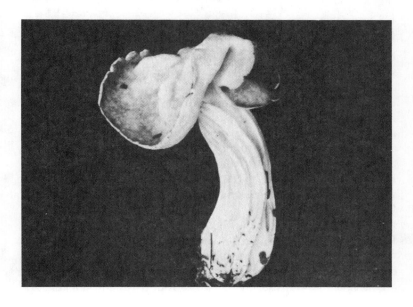

FLEXIBLE HELVELLA
Helvella elastica Bull. ex St. Amans

FRUITING BODIES: Solitary to gregarious; *pileus* grayish tan to medium brown, saddle-shaped to mitral; attached to stipe at apex only, up to 3 cm wide; *stipe* white, smooth, solid or stuffed up to 1 cm wide and 8 cm long.

MICROCHARACTERS: *Spores* 18-22 x 10-14 um, oblong-elliptic, smooth at maturity, coarsely warted when young, with one central oil drop. *Asci* 260-300 x 17-19 um, J-. *Paraphyses* clavate, occasionally branched, filled with oil drops, 6-10 um broad at the apex. *Medullary excipulum* of textura intricata, hyphae 3-7 um broad, nearly colorless. *Ectal excipulum* of textura globulosa or textura angularis, inner cells up to 50 um broad, giving rise to a densely packed tomentum of short chains of cells up to 80 um long.

HABITAT-DISTRIBUTION: Occurring throughout the growing season in coniferous woods on soil, duff or moss, common in western U.S.

DESCRIPTIONS: refs. 16, 31, 37, 43.

REMARKS: The small size, sparse fruiting, and fragile texture would discourage table use. Smith (39) indicates it is reportedly edible. *Helvella albipes* appears to be closely related to *H. elastica* and is said to differ (36) by its two to four lobes and thicker stipe.

ELFIN SADDLE

Helvella lacunosa Afz. ex Fr.

FRUITING BODIES: Scattered to gregarious; *pileus* from whitish to gray to nearly black, saddle-shaped to convolute, margin attached to stipe at several points, up to 5 cm wide; *stipe* whitish to medium gray, ribbed, and channeled within, up to 10 cm long and 1 cm wide.
MICROCHARACTERS: *Spores* 12-21 x 12-14 um oblong-elliptic with slightly truncate ends, smooth, with one large central oil drop. *Asci* 250-300 x 14-20 um, J-. *Paraphyses* clavate, occasionally branched, the tips up to 9 um wide, brownish in mass. *Medullary excipulum* of textura intricata, the hyphae up to 7 um wide. *Ectal excipulum* of textura angularis, thin, not well differentiated, outermost cells catenulate.
HABITAT-DISTRIBUTION: Widespread, common in western U.S. in conifer forests, and mixed conifer and deciduous woods on soil and rotting wood, late summer and fall.
DESCRIPTIONS: refs. 6, 31, 37.
REMARKS: A very variable species as to color, but is readily distinguished by the attachment of pileus margin to stipe at several points. Both Smith (39) and Miller (28) report it as edible. I have not tried it. See cautions mentioned under *H. crispa.*

WHITE FOOTED ELF-CUP
Helvella leucomelaena (Pers.) Nannf.

FRUITING BODIES: Gregarious to cespitose; cupulate, short stipitate, stipe usually subterranean, margin delicately lobed and splitting, up to 5 cm wide; *hymenium* dark gray to dark grayish brown, smooth; *exterior* about concolorous with the hymenium near margin, gradually shading out to nearly white at the stipe; *stipe* white, with rounded ribs that do not extend up onto the cup, channeled within.

MICROCHARACTERS: *Spores* 20-24 x 10-13 um, elliptical, smooth, with a single large central oil drop. *Asci* 250-300 x 13-14 um, J-. *Paraphyses* occasionally branched, clavate, the apices up to 7 um wide. *Medullary excipulum* of textura intricata, the hyphae 3-5 um wide. *Ectal excipulum* of textura globulosa and textura angularis, inner cells 15-30 um, outer cells 20-40 x 12-12 um, catenulate.

HABITAT-DISTRIBUTION: Spring and summer, western U.S., often associated with grassy areas in coniferous woods, on soil, skid roads, etc., common.

DESCRIPTIONS: refs. 16, 31, 43.

REMARKS: Edibility untested. This distinctive fungus has long gone under the name of *Paxina leucomelas* in North America.

BIG FOOTED HELVELLA
Helvella macropus (Pers. ex Fr.) Karst.

FRUITING BODIES: Solitary to gregarious; cupulate with a long narrow stalk, margin incurved at first, up to 5 cm wide; *hymenium* gray-brown, smooth; *exterior* concolorous or paler than hymenium, granular to fibrillose scaly; *stipe* terete, solid, up to 7 cm long concolorous with exterior, paler below, scaly granulose as the exterior of cup.

MICROCHARACTERS: *Spores* 20-25 x 10-12 um, elliptic-fusoid, smooth to verrucose-rugulose, especially when young, with one large central oil drop and two small polar ones. *Asci* 225-275 x 14-16 um, J-. *Paraphyses* clavate, septate, up to 12 um wide at apex. *Medullary excipulum* of textura intricata, hyphae mostly 3-6 um wide, but occasionally up to 30 um wide. *Ectal excipulum* of textura globulosa to textura angularis the outer cells giving rise to fascicled tufts up to 200-400 um long.

HABITAT-DISTRIBUTION: Summer and fall on soil or rotting wood in conifer and mixed forests, widespread worldwide, but not very often encountered.

DESCRIPTIONS: refs. 6, 16, 43.

REMARKS: Edibility not tested. *Helvella corium* and *H. villosa* look very much like *H. macropus.* They may be separated primarily by their darker blackish color and by the microscopic characters given in the key.

QUELET'S HELVELLA
Helvella queletii Bres.

FRUITING BODIES: Scattered to gregarious; cupulate to plane or trilobate, short-stipitate up to 8 cm wide; *hymenium* brown to gray brown; *exterior* paler than hymenium, smooth or delicately granulose; *stipe* tan up to 8 cm long, ribbed, but ribs not extending over lower surface of cup, smooth or delicately granulose.

MICROCHARACTERS: *Spores* 20-24 x 11-14 um, elliptical, smooth, with a single large central oil drop. *Asci* 220-310 x 13-17 um, J-. *Paraphyses* slightly clavate up to 8 um wide at apex. *Medullary excipulum* of textura intricata, the hyphae 3-5 um wide. *Ectal excipulum* of textura angularis to textura globulosa, inner cells 15-24 um wide, outer cells giving rise to a tomentum of clavate hairs up to 80 x 25 um which form the granules on exterior and stipe.

HABITAT-DISTRIBUTION: Late spring and summer, on soil or duff, widespread throughout the western U.S. but not common.

DESCRIPTIONS: refs. 16, 31, 43.

REMARKS: Edibility untested. This species intergrades with *H. acetabulum* but never exhibits the prominent ribbing on the underside of the cup.

HAIRY HELVELLA
Helvella villosa (O. Kuntz) Dissing and Nannfeldt

FRUITING BODIES: Solitary or aggregated; cupulate to flat with a long narrow stalk, margin straight, curved up to curved down; 2-4 cm wide; *hymenium* dark gray, smooth; *exterior* the same color or paler, dull from the abundant short velvety hairs; smooth; *stipe* colored the same as exterior of cup but pale or whitish at base, slightly wider at base or nearly cylindrical, solid, surface smooth, delicately velvety hairy, up to 5-6 cm long, 4-5 mm wide at apex, and abruptly flaring into the apical cupulate area; context white, soft.

MICROCHARACTERS: *Spores* 16-19 x 13-14 um, broadly ellipsoid, with a prominent central oil drop and one or more smaller drops, smooth, hyaline. *Asci* 210-300 x 15 um, J-. *Paraphyses* clavate to cylindric, up to 7 cm wide at apex, septate, pale grayish tan. *Medullary excipulum* of textura intricata, the hyphae 6-12 um wide, hyaline or with scattered gray brown pigment. *Ectal excipulum* of textura globulosa to textura angularis; many globose to elongate cells forming short chains, globose cells 15-20 um wide, some pigmented.

HABITAT-DISTRIBUTION: Fall, on soil in conifer forests usually along streams and shaded flood plains.

DESCRIPTIONS: refs. 11, 16, 43.

REMARKS: Edibility not tested. The hairy nature of the exterior and stipe may be easily overlooked if not viewed with a lens. This is the only species in the cupulate *Helvella* group in which the cup is truly dark gray, not black or dark brown. Compare with *H. corium* and *H. macropus* if uncertain of the fruiting body color.

Mushroom habitat: Check a burn in the spring following a forest fire. Morels and other Discomycetes will often fruit abundantly in such areas. Scene is of Piccolo Creek Burn, Latah County, Idaho.

TRICORNERED ELF-SADDLE
Paxina recurvum Snyder

FRUITING BODIES: Scattered to gregarious; stipitate pileate; *pileus* yellow brown, turned down on three or four sides to give an angular appearance, up to 10 cm wide; *exterior* or reverse side pale; *stipe* whitish, round, smooth, lacunose within, up to 4 cm long and 2 cm wide.

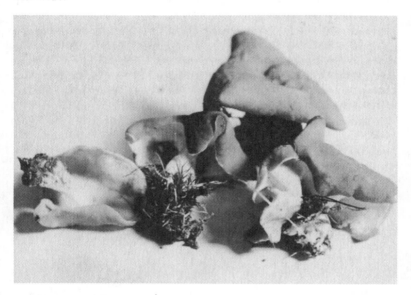

MICROCHARACTERS: *Spores* 12-14 x 7-10 um, elliptical to oval, minutely verrucose, the verrucae noncyanophilic, usually with two oil drops. *Asci* 250-300 x 13-15 um, J-. *Paraphyses* clavate stout, the base 5-6 um wide, the apex up to 10 um wide. *Medullary excipulum* of loosely interwoven textura intricata. *Ectal excipulum* hardly discernable.

HABITAT-DISTRIBUTION: Northern Idaho, British Columbia, Montana and Washington, spring and early summer, on soil or duff in coniferous forests, not common.

DESCRIPTIONS: refs. 31, 37, 41.

REMARKS: Edibility untested. This species is very close to *Peziza melaleucoides* on the basis of microscopic characters. The angular cap and non-lignicolous habit distinguish it.

ROOTING FAIRY CUP
Rhizina undulata Fr.

FRUITING BODIES: Single to gregarious; flat, or irregularly lobed, attached to substrate over entire pale colored lower side by numerous whitish to yellowish rhizoids, up to 6 cm wide; *hymenium* dark purple brown to blackish, margin pale yellow just like the underside, undulate and irregular, fleshy-tough.

MICROCHARACTERS: *Spores* 30-40 x 8-11 um, fusoid, apiculate, minutely verrucose at maturity, with one or two oil drops. *Asci* 400-500 x 11-14 um, J-. *Paraphyses* slightly clavate, tips encrusted with amorphous brown material, intermixed with tubular setae, thin walled, brown, aseptate and parallel-sided, tapering to a blunt point, 7-11 um wide. *Medullary excipulum* of densely interwoven textura intricata up to 1 mm thick, the hyphae up to 8-10 um thick. *Ectal excipulum* of textura globulosa, the cells up to 50 um wide.

HABITAT-DISTRIBUTION: Widespread, not common, terrestrial on burned soil or conifer debris or parasitic on conifer seedlings. Northern U.S. spring and summer.

DESCRIPTIONS: refs. 6, 31, 37.

REMARKS: Edibility untested. A rather distinctive species with a crust-like habit and numerous rhizoids.

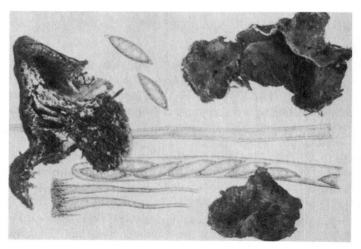

Photograph taken from Seaver (37).

84

7. Pyronemataceae

This large family includes many small, inconspicuous and rarely encountered fungi. They are usually not collected for the table, but their brilliant red and orange colored apothecia never cease to attract the attention of the mushroom collector. Those which are 1 cm wide or larger are keyed in key #1 p. 86 and illustrated and described. Those less than 1 cm wide are listed and keyed in key #2 p. 88 but are neither illustrated nor described. The dull or pale colored species are easily confused with the members of the Pezizaceae. If a microscope and Melzer's reagent are used, it will be found that the asci in this group are iodine negative (J-), whereas those in the Pezizaceae are iodine positive (J+). This one feature alone will serve to separate the two groups.

Diagnostic Description
Apothecia fleshy to brittle, cupulate to discoid, usually not stalked, at least not pileate; asci not bluing with iodine, typically cylindric with apical operculum; ascospores thin walled, smooth or rough, hyaline, or if brown, then pigment deposited from inner side, without external polar granules, uninucleate; paraphyses not anastomosing; on various substrata, dung, soil, wood, etc.

LIST OF SPECIES
1. *Aleuria aurantia* (Fr.) Fuckel
2. *Anthracobia macrocystis* (Cke.) Boud.
3. *Anthracobia melaloma* (Fr.) Boud.
4. *Caloscypha fulgens* (Pers.) Boud.
5. *Cheilymenia coprinaria* (Cke.) Boud.
6. *Cheilymenia crucipila* (Cke. & Phill.) LeGal
7. *Cheilymenia stercoraria* (Fr.) Boud.
8. *Cheilymenia theloboloides* (Fr.) Boud.
9. *Coprobia granulata* (Fr.) Boud.
10. *Coprotus ochraceus* (Cr. & Cr.) Larsen
11. *Geopora cooperi* Harkness
12. *Geopyxis carbonaria* (Pers.) Sacc.
13. *Geopyxis vulcanalis* (Pk.) Sacc.
14. *Humaria hemisphaerica* (Wigg. Ex Fr.) Fuckel
15. *Lamprospora crec' hqueraultii* (Cr.) Boud.
16. *Lamprospora spinulosa* Seaver
17. *Lasiobolus ciliatus* (Fr.) Boud.
18. *Leucoscypha rutilans* (Fr.) Dennis & Rifai

19. *Meladina lechitina* (Cke.) Svrcek
20. *Melastiza chateri* (W.G. Smith) Boud.
21. *Octospora leucoloma* Hedwig ex S.F. Gray
22. *Otidea abietina* (Fr.) Fuckel
23. *Otidea alutacea* (Pers.) Massee
24. *Otidea concinna* (Pers.) Sacc.
25. *Otidea grandis* (Pers.) Rehm.
26. *Otidea leporina* (Fr.) Fuckel
27. *Otidea onotica* (Fr.) Fuckel
28. *Otidea smithii* Kanouse
29. *Psilopezia nummularia* Berk.
30. *Pulvinula archeri* (Berk.) Rifai
31. *Pyronema omphaloides* (St. Amans) Fuckel
32. *Scutellinia erinaceus* (Schw.) Kuntze
33. *Scutellinia scutellata* (Fr.) Lambotte
34. *Scutellinia umbrarum* (Fr.) Lamb.
35. *Tarzetta cupularis* (L. ex Fr.) Lamb.
36. *Tricharina gilva* (Boud.) Eckblad
37. *Trichophaea abundans* (Karst.) Boud.
38. *Trichophaea boudierii* Grel.

Pyrenomycetaceae Key #1
Apothecia Large, 1cm or Larger

1a Apothecia above ground 2
1b Apothecia hypogeous .... *Geopora cooperi*
2a Margin of apothecia adorned with stiff blackish brown or pale brown hairs 3
2b Margin of apothecium without stiff bristle-like brown to black hairs 7
3a Fruiting on dung; hymenium orange, hairs dark brown, thick walled, forked and rooting, basally unbranched; spores 17-22 x 10-12 um, elliptical, smooth, eguttulate .... *Cheilymenia coprinaria*
3b Fruiting on soil or rotting wood 4
4a Hymenium grayish or whitish................................ 5
4b Hymenium orange or red 6
5a Hymenium grayish, exterior pale brown, hairs sparingly septate; ascospores broadly fusoid, with numerous oil drops 23-25 x 12-13 um .... *Trichophaea boudieri*
5b Hymenium whitish; exterior with septate, brown, closely septate bristle-like hairs; ascospores elliptical, smooth to verrucose, with 2-3 guttulae, hyaline, 10-11 x 20-22 um; on soil or rotting wood .... *Humaria hemispherica*
6a Hymenium red; apothecia cupulate to discoid, up to 1 cm wide; marginal hairs long, up to 1 mm, scattered to abundant, thick-walled,

86

septate, dark brown, unbranched, tapering to a point, short hairs usually also present; ascospores oval, with one or more oil drops, rough with small anastomosing warts, 11-14 x 17-19 um; on rotting wood or soil . . . . *Scutellinia scutellata*

6b Hymenium orange or orange red; apothecia cupulate to discoid, up to 2 cm wide; marginal rooting hairs 150-700 um long, septate, pointed, dark brown; ascospores elliptical, 1-2 guttulate, warty with scattered rounded warts 14-16 x 20-24 um; on soil, rarely on wood . . . .
. . . . *Scutellinia umbrarum*

7a Margin of apothecium or hymenium with olive-green tinge; apothecium up to 6 cm wide, deep cupulate at first, later repand and occasionally split or one sided, sessile or with short yellowish stipe; hymenium bright orange at times spotted green; ascospores smooth, hyaline, spherical with granular contents, 6-8 um wide; paraphyses filled with orange granules, filiform; on duff, often shortly after snows recede in spring
. . . . *Caloscypha fulgens*

7b Margin of apothecium or hymenium without olive-green tinge 8

8a Apothecia one sided ear-or spoon-shaped 12

8b Apothecia cupulate to repand 9

9a Hymenium bright orange; apothecia cupulate to soon discoid, up to 6 cm wide, fragile; ascospores hyaline, elliptical, biguttulate, reticulate at maturity, 9-11 x 17-22 um; paraphyses with orange granules that turn green in iodine; on soil, fall . . . . *Aleuria aurantia*

9b Hymenium pale, yellow, to gray brown or reddish brown 10

10a Ascospores biguttulate; apothecia subsessile, cupulate, grayish tan, up to 2 cm wide; margin crenate; exterior pustulate; ascospores elliptical, smooth, biguttulate, hyaline, 10-12 x 20-23 um; paraphyses occasionally bent, clavate, 5 um wide at apex; on soil . . . .
. . . . *Tarzetta cupularis*

10b Ascospores biguttulate; apothecia subsessile, cupulate, grayish tan, up

11a Hymenium brick red; paraphyses with orange brown granules; tips not forked or lobed; apothecia deep cupulate, stipitate, up to 1 cm wide; margin crenate, white; stipe whitish, thin; ascospores elliptical, smooth, 7-9 x 13-18 um; on burned soil . . . . . . *Geopyxis carbonaria*

11b Hymenium pale yellow; paraphyses without orange brown granules, tips often lobed, forked or bent; apothecia short stipitate, cupulate to nearly discoid, up to 1 cm wide; margin crenate, white; ascospores smooth, hyaline, elliptical, 8-11 x 15-21 um; on unburned or burned soil . . . . *Geopyxis vulcanalis*

12a Apothecia bright yellow with rosy blush to hymenium, spoon or ear-shaped, substipitate; margin involute; base white tomentose; ascospores hyaline or faint yellow, ellipsoid, smooth, biguttulate, 12-14 x 6-7 um; paraphyses filiform, usually hooked, base forked at times; in conifer duff, summer and fall . . . . *Otidea onotica*

12b Apothecia yellow, yellow brown, or vinaceous to dark brown, without rosy flush to hymenium 13

13a Apothecia typically spoon or ear-shaped 17
13b Apothecia with truncate apex, not ear-shaped..................14
14a Apothecia clear yellow, solitary to cespitose, 3 cm high by 4 cm wide; stem-like base whitish tomentose below; ascospores smooth, biguttulate, slightly yellow, ellipsoid, 5-6 x 10-12 um; paraphyses hooked or bent at apex, forked below; on the ground .... *Otidea concinna*
14b Apothecia avellaneous or dark to medium brown 15
15a Apothecia with light yellow brown exterior, cespitose, up to 6 cm high, 4 cm wide, fragile; hymenium avellaneous; ascospores smooth, narrowly ellipsoid, biguttulate, slightly yellowish, 7-9 x 14-16 um; paraphyses occasionally branched below, apex hooked; on coniferous duff .... *Otidea alutacea*
15b Apothecia medium to dark brown...........................16
16a Apothecia medium brown, paraphyses occasionally with short lateral branches or notches usually on underside of bent portion; apothecia liver brown, split on one side, truncate; ascospores 18-20 x 10-12 um, elliptical smooth, faint yellowish, biguttulate; on the ground .... *Otidea abietina*
16b Apothecia dark brown, solitary to cespitose, stipitate, up to 2 cm high, 4 cm wide, leathery; hymenium vinaceous fawn, frequently with red-orange patches; stipe thick up to 1 cm long, yellowish; ascospores long ellipsoid to subfusoid, biguttulate, outer wall smooth, inner wall in age slightly roughened 6-7 x 14-17 um; paraphyses hooked, forked below; on the ground .... *Otidea grandis*
17a Apothecia deep vinaceous brown, large, up to 8 cm high, arising from a large solid foot like base, fragile; ascospores smooth, hyaline or faintly yellowish, biguttulate, narrowly ellipsoid, 6-8 x 10-15 um; paraphyses with hooked apices, at times tips with irregular protuberances; on conifer duff .... *Otidea smithii*
17b Apothecia dull yellowish brown, up to 4 cm high; exterior cinnamon rufous; hymenium yellowish brown; base narrowed to a cream white stipe; ascospores smooth, biguttulate, ellipsoid, 6-8 x 12-14 um; paraphyses filiform, tips hooked; on conifer duff .... *Otidea leporina*

Pyrenomycetaceae Key #2
Apothecia Small, 1 cm or less

5b(3b) Hymenium yellow, dull orange, brownish yellow, grayish, etc 16

6a(5a) Apothecia on dead *Abies* branches that usually still retain many old needles, often near snowbanks in spring . . . . *Pithya vulgaris* of Sarcosyphaceae

6b(5a) Apothecia not on *Abies* branches which still retain the old needles 7

7a(6b) Growing on submerged or very wet wood, margin of apothecium of angular or globose cells; apothecia discoid adhering closely to substrate, surrounded by radiating white mycelium, hymenium bright orange, blackish when dry; ascospores 20-24 x 12-13 um, smooth or minutely sculptured . . . . *Meladina lechitina*

7b(6b) Not on submerged or very wet wood 8

8a(7b) Growing on the ground among mosses, apothecia small, not exceeding 1 mm, pale orange; ascospores 18-50 um, with a single oil drop, spiny at maturity the spines short, blunt . . . . *Lamprospora spinulosa*

8b(7b) Growing on soil, not among mosses 9

9a(8b) Fruiting in the autumn; apothecia typically occurring in large numbers, large, mostly above 5 mm wide to well above 1 cm wide . . . . *Aleuria aurantia*

9b(8b) Fruiting vernal; apothecia smaller 2-5 mm wide 10

10a(9b) Exterior with short pale hairs; ascospores 15-22 um; globose, with many oil drops, spiny at maturity, the spines long, up to 3 um, sharp and blunt pointed; on soil . . . . *Lamprospora crec' hqueraultii*

10b(9b) Exterior with clumps of hairs, hairs short, easily overlooked, obtuse, thin walled and light brown; ascospores 19-24 x 9-14 um, elliptical with two oil drops; rough with cyanophilous reticulum; paraphyses with orange droplets which turn green in iodine; on wet soil . . . . *Melastiza chateri*

11a(1b) Apothecia hairy 12

11b(1b) Apothecia glabrous, pustulate, scurfy, etc.; not hairy 15

12a(11a) Apothecia with stiff hyaline hairs, hairs arising from excipulum near the base and often exceeding the apothecia in length, thickwalled, aseptate, widest above the base, tapering to a point; apothecia pale orange, soft and fleshy; spores 21-28 x 9-14 um, elliptical, smooth; on dung . . . . *Lasiobolus ciliatus*

12b(11a) Hairs brown, yellow-brown or reddish-brown 13

13a(12b) Cruciate hairs present; hairs of two types, marginal bristle-like thickwalled, brown, septate, rooting hairs, intermingled below with stellate, yellow to medium brown, 2-5 pointed hairs; apothecia red-orange; ascospores 19-23 x 10-13 um, elliptical, smooth; on dung . . . . *Cheilymenia stercorarea*

13b(12b) Cruciate hairs absent 14

14a(13b) Apothecia yellow, nearly smooth but with few scattered unbranched, straight, yellow brown hairs; ascospores 17-18 x 9-11 um, elliptical; excipulum one layered of textura globulosa and textura angularis; on dung, soil or duff . . . . *Cheilymenia theleboloides*

14b(13b) Apothecia reddish orange .... *Cheilymenia coprinaria*

15a(11b) Apothecia, sessile, hymenium orange, exterior paler, coarsely granulose; ascospores 15-18 x 7-10 um, eguttulate, elliptical, smooth; paraphyses with orange granules which turn green in iodine; on dung .... *Coprobia granulata*

15b(11b) Apothecia pale yellow to orange, subglobose, externally smooth up to 1 mm wide; excipulum of textura angularis to globulosa, cyanophilic; ascospores 14-18 x 19-11 um, broadly ellipsoid smooth, hyaline to slightly yellowish, with one deBary bubble; on dung .... *Coprotus ochraceous*

16a(5b) Margin scalloped or regularly notched 17

16b(5b) Margin even, or irregularly notched; hymenium reddish brown to brown; ascospores 20-25 x 12 um, ellipsoid, with one or two oil drops with perispore which loosens in hot lactic acid; on rotting wood .... *Psilopezia nummularia*

17a(16a) Apothecium discoid; hymenium dull orange, crenate; ascospores 18-24 x 13-15 um, broadly elliptical, smooth, 1-3 oil drops; margin with long clavate cells; on soil among mosses .... *Ostospora leucoloma*

17b(16a) Apothecium deep cupulate, grayish yellow, crenate; ascospores with two large oil drops .... *Tarzetta cupularis*

18a(4b) Hymenium pale gray; apothecia with brown bristle-like hairs, mostly confined to margin; ascospores 23-25 x 12-13 um, broadly fusoid, with many oil drops; on soil .... *Trichophaea boudieri*

18b(4b) Hymenium orange, yellow, whitish, etc. 19

19a(18b) Hymenium bright orange 19

19b(18b) Hymenium yellow, whitish etc., not orange 21

20a(19a) Ascospores finely warty, cruciate hairs present; apothecia bright orange to orange red; hairs of two types, marginal hairs unbranched up to 500 um long, 2-5 septate tapering to a rounded point, and rooting in the medullary excipulum, the lower (cruciate) hairs mostly branched with 2-5 arms and shorter up to 150 um long; ascospores 15-18 x 8-9 um, exospore separating in hot lactic acid, sometimes with pseudoapiculi; on soil .... *Cheilymenia crucipila*

20b(19a) Ascospores rough with cyanophilous reticulum; hymenium bright orange; hairs short, easily overlooked, obtuse, thin walled and light brown; ascospores 19-24 x 9-14 um, elliptical with two oil drops; paraphyses with orange droplets which turn green in iodine; on wet soil .... *Melastiza chateri* and *Leucoscypha rutilans*

21a(19b) Hymenium whitish, exterior densely hairy with brown hairs, tapering to sharp point; ascospores 20-22 x 10-11 um, elliptical, smooth or slightly verrucose, with 2-3 oil drops............. .... *Humaria hemispherica*

21b(19b) Hymenium yellow, exterior nearly smooth, but with few scattered unbranched, straight yellow brown hairs; ascospores 17-18 x 9-11 um, elliptical, smooth, without oil drops; excipulum of textura globulosa and textura angularis; on dung, soil or duff

90

.... *Cheilymenia theleboloides*
22a(2a) Apothecia with brown to reddish brown hairs 23
22b(2a) Apothecia not hairy, but may be subtended by a hairy subiculum
.... 25
23a(22a) Ascospores eguttulate; apothecia with reddish-brown hairs, sessile; hymenium ochraceous orange to grayish-orange; ascospores 14-19 x 9-11 um, elliptical, smooth, hyaline; on burned soil, ashes etc. .... *Tricharina gilva*
23b(22a) Ascospores with 2 oil drops 24
24a(23b) Hymenium yellow brown; apothecia covered with short brown septate obtuse hairs, mostly aggregated into fascicles; ascospores 15-21 x 8-10 um, elliptical; on burned soil ....
.... *Anthracobia melaloma*
24b(23b) Hymenium red; apothecia clothed with obtuse, brown, septate hairs; ascospores 14-16 x 7-9 um, elliptical; on burned soil ....
.... *Anthracobia macrocystis*
25a(22b) Apothecia small 1-2 mm wide confluent and forming an extensive mass on a subiculum, orange to reddish orange; ascospores 11-15 x 6-8 um; ellipsoid; often on burned soil, wood, plaster, etc....
.... *Pyronema omphaloides*
25b(22b) Apothecia larger not formed on an extensive subiculum 26
26a(25b) Apothecia deep cupulate, margin crenate whitish; ascospores elliptical eguttulate......................................
.... *Geopyxis vulcanalis* and *Geopyxis carbonaria*
26b(25b) Apothecia pulvinate; paraphyses narrow, filiform, branched, with curved tips; hymenium pale to bright orange; ascospores 7-9 um wide, globose, contents granular, smooth; on burned soil
.... *Pulvinula archeri*

Foray activity: Mushroom collectors sorting out a day's find at Laird Park, a popular collecting area in Northern Idaho.

91

ORANGE PEEL FUNGUS
Aleuria aurantia (Fr.) Fuckel

FRUITING BODIES: Single or aggregated in compact clusters; cup-shaped to irregularly cupulate with one side infolded or split; glabrous, thin and fragile, up to 4-6 cm wide, *hymenium* bright orange to reddish orange; *outer surface* paler, pruinose, mealy roughened, not hairy, sessile, *mycelium* white.

MICROCHARACTERS: *Spores* 17-22 x 9-11 um, elliptical, with two oil drops, hyaline, roughened by a reticulum at maturity which may disappear in KOH preparations, smooth at first. *Asci* 215-230 x 12-13 um, cylindric, not bluing with iodine, not projecting. *Paraphyses* slightly clavate, 7-10 um wide at tips, with numerous orange granules which turn green in Melzer's solution. *Hypothecium* textura angularis. *Medullary excipulum* thin, of globose cells, forming short hyaline hairs and pustules.

HABITAT-DISTRIBUTION: On mineral soil along roads and banks, widespread and common in western U.S. and Canada, fall.

DESCRIPTIONS: refs. 6, 31, 37.

REMARKS: Our most beautiful and commonly encountered fall cup fungus. Those collecting in disturbed areas, road cuts and the like, will find it on bare ground. Edible but too fragile to collect for the table.

SPRING ORANGE PEEL FUNGUS
Caloscypha fulgens (Pers.) Boud.

FRUITING BODIES: Single to gregarious; deep cupulate to otidioid, often split or compressed at one side, margin incurved when young expanding to nearly flat, short stipitate or sessile up to 6 cm wide; *hymenium* bright orange, stained uniformly olive green or only so on the margin when young, smooth, or wrinkled; *exterior* darker from olive hue, smooth to pruinose, not hairy.

MICROCHARACTERS: *Spores* 6-8 um in diameter, globose, smooth, eguttulate but with granular contents. *Asci* 80-100 x 7-8 um, cylindric, not projecting, J-. *Paraphyses* filiform, septate with numerous yellow-orange granules. *Hypothecium* of textura intricata. *Medullary excipulum* of loose textura intricata with oily material in the hyphae. *Ectal excipulum* of textura angularis with few outer hairs at times and nearly textura epidermoidea at times. The conidial stage is the seed pathogen *Geniculodendron pyriforme* (32).
HABITAT-DISTRIBUTION: Widespread during spring in Rocky Mountains and the Pacific Northwest, on soil or duff under conifers shortly after snows recede.
DESCRIPTIONS: refs. 31, 37, 38.
REMARKS: Our most common early discomycete, and a conspicuous element of our early spring mushroom flora. The hymenium is always bright orange but the degree of olive color varies. An albino form of this fungus in which the orange coloring matter was absent was encountered by Dr. J. Rogers near Laird Park

in northern Idaho in 1978. The pigment which stains the normal apothecia olive-green, was instead blue in the otherwise pure white albino.

DUNG LOVING ELF CUP
Cheilymenia coprinaria (Cke.) Boud.

FRUITING BODIES: Single to gregarious; discoid, sessile, margin hairy up to 1 cm wide; *hymenium* bright yellow orange, smooth; *exterior* paler, hairy with brown, shiny pointed bristle-like hairs.

MICROCHARACTERS: *Spores* 17-22 x 10-12 um, ellipsoid, smooth, eguttulate. *Asci* 210-300 x 12-15 um, J-. *Paraphyses* slightly clavate, containing orange granules, tips up to 7 um wide. *Medullary excipulum* of small irregular cells. *Ectal excipulum* of large, ellipsoidal cells 40-100 x 20-50 um, with the long axis perpendicular to the exterior.

HABITAT-DISTRIBUTION: Worldwide, common on dung during the growing season.

DESCRIPTIONS: refs. 6, 31, 37.

REMARKS: Edibility untested. Many other small orange to reddish orange cup fungi occur on dung. These are seldom of interest to the average collector but additional information may be obtained from

Seaver (37) or Dennis (6). *Cheilymenia crucipila* of similar habit grows on the ground among moss.

COOPER'S TRUFFLE
Geopora cooperi Harkness

FRUITING BODIES: Subterranean, globose; surface lobed from internal folds, and fuzzy from brown hairs, yellow brown to dark brown; external dense mycelial mass eventually breaking down to expose the paler whitish folded hymenium.

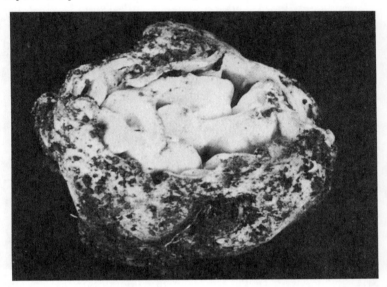

MICROCHARACTERS: *Spores* 20-27 x 13-17 um, broadly oval-elliptical, smooth, uniguttulate, hyaline, violently discharged. *Asci* 180-280 x 15-22 um, operculate, 8 spored, J-. *Paraphyses* slightly clavate, hyaline, septate up to 10 um wide at apex. *Medullary excipulum* of textura intricata, the hyphae 6-10 um wide. *Ectal excipulum* of textura angularis, the cells up to 50 um wide and yellow brown.

HABITAT-DISTRIBUTION: Widespread in Pacific Northwest under conifers in subalpine areas, common, but rarely collected except by truffle collectors.

DESCRIPTION: ref. 3.

REMARKS: A remarkable fungus, first thought to be a truffle until

Burdsall (3) showed the spores are discharged violently just as in other Pezizales. Edibility unknown. *Hydnotya cerebreformis* of the *Tuberales* has similar convoluted internal tissue, but differs in many important microscopic characters.

CHARCOAL LOVING ELF CUP
Geopyxis carbonaria (Pers.) Sacc.

FRUITING BODIES: Gregarious; deep cupulate, stipitate, margin crenate, white, up to 1 cm wide; *hymenium* brick red, smooth; *exterior* dull yellowish, pustulate or nearly smooth; *stipe* whitish, short, expanding abruptly into the cup.

MICROCHARACTERS: *Spores* 13-18 x 7-9 um, elliptical, smooth, eguttulate. *Asci* 190-225 x 9-10 um, J-. *Paraphyses* slightly clavate, unbranched, with irregular orange-brown granules, tips up to 5 um wide not forked or lobed. *Hypothecium* of densely packed small irregular cells. *Medullary excipulum* of narrow, hyaline, densely interwoven textura intricata. *Ectal excipulum* of textura angularis.
HABITAT-DISTRIBUTION: Widespread on burned soil or charcoal in the spring and throughout the growing season.
DESCRIPTIONS: refs. 6, 31, 37.
REMARKS: Edibility untested. *G. vulcanalis* with which it may be confused has a yellow apothecium, and paraphyses lacking orange brown granules. *Tarzetta cupularis* of similar habitat is readily distinguished by its biguttulate ascospores.

VULCAN ELF CUP
Geopyxis vulcanalis (Pk.) Sacc.

FRUITING BODIES: Scattered to gregarious; cupulate, to nearly discoid, short stipitate, margin inturned, white, crenate, up to 1 cm wide; *hymenium* pale yellow; *exterior* paler, pruinose, glabrescent; *stipe* whitish usually buried in substrate.

MICROCHARACTERS: *Spores* 15-21 x 8-11 um, elliptical, smooth, hyaline, eguttulate. *Asci* 250-270 x 13-16 um, J-. *Paraphyses* filiform, hyaline, without orange brown granules, tips bent, forked or lobed, up to 3 um wide. *Hypothecium* of hyaline densely interwoven textura intricata, the hyphae up to 4 um wide. *Medullary excipulum* two layered, the upper of narrow hyphae 3-4 um wide, and the lower of larger more loosely interwoven hyphae 4-8 um wide. *Ectal excipulum* of textura angularis.

HABITAT-DISTRIBUTION: Common on soil (rarely burned) or conifer litter, or among moss in spring, summer and fall, widespread, North America.

DESCRIPTIONS: refs. 31, 37.

REMARKS: Edibility not tested. *Tarzetta cupularis* is similar but has biguttulate ascospores.

TRUNCATE EAR FUNGUS
Otidea alutacea (Pers.) Massee

FRUITING BODIES: Gregarious to densely clustered; fragile, deep cupulate, apex truncate, short stipitate, up to 6 cm high and 4 cm wide, glabrous; *hymenium* light grayish brown to light brown; *exterior* light yellow brown.

MICROCHARACTERS: *Spores* 14-16 x 7-9 um, elliptical, smooth, hyaline, biguttulate. *Asci* 160-350 x 10-14 um. *Paraphyses* filiform, septate, hooked at apices, occasionally branched, up to 4 um wide at apex. *Medullary excipulum* of textura intricata. *Ectal excipulum* of textura angularis, the outer cells catenulate forming a few short chains of hyaline cells.

HABITAT-DISTRIBUTION: On the ground in the fall in coniferous forests of the Pacific Northwest.

DESCRIPTIONS: refs. 14, 31.

REMARKS: Edibility untested. The truncate apex of the fruiting body and the hooked paraphyses are diagnostic features of this species. The variety *O. alutacea* var. *microspora* which also occurs here is distinguished by the smaller spores 9-10 x 5.5-6.5 um, paler yellow apothecia and a thick, tough, medullary excipulum.

EAR FUNGUS
Otidea leporina (Fr.) Fuckel

FRUITING BODIES: Gregarious to cespitose; spoon-shaped or elongated ear-shaped, sessile to short-stipitate, up to 4 cm high; *hymenium* yellowish to yellow brown, smooth; *exterior* colored the same or slightly paler, delicately pustulate.

MICROCHARACTERS: *Spores* 10-14 x 6-8 um, elliptical to oblong-elliptical, smooth, hyaline, biguttulate or at times uniguttulate. *Asci* 140-190 x 9-12 um. *Paraphyses* filiform, branched, curved at apex, up to 4.5 um wide. *Medullary excipulum* of hyaline densely interwoven textura intricata. *Ectal excipulum* of textura angularis, the cells yellowish, catenulate, and forming minute pustules encrusted with yellow-brown granules which are somewhat soluble in KOH.

HABITAT-DISTRIBUTION: Common in the late summer and fall on the ground in conifer forests, especially Douglas Fir, in the Pacific

Northwest.

DESCRIPTIONS: refs. 14, 31.

REMARKS: Edibility untested. *Otidea concinna* is very similar if not identical. For collections which resemble *Otidea leporina* but have a narrowly elongated or pointed apex and large size see *Wynnella silvicola* of the Helvellaceae. The latter is distinguished readily on spore characters.

EAR FUNGUS
Otidea smithii Kanouse

FRUITING BODIES: Solitary to cespitose; ear-shaped to truncate, fragile, stipitate, up to 8 cm high; *hymenium* medium yellow brown; *exterior* darker with a grayish or vinaceous cast; *stipe* concolorous or tan, extending into substrate to form a foot like base of soil and mycelium.

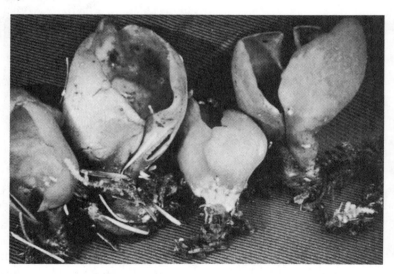

MICROCHARACTERS: *Spores* 12-15 x 6-8 um, elliptical, smooth, hyaline, biguttulate. *Asci* 150-200 x 11-13 um, J-. *Paraphyses* filiform, hooked at apex, unbranched, sparingly septate, often with protuberances. *Hypothecium* of hyaline densely interwoven textura intricata. *Medullary excipulum* of more loosely interwoven textura intricata, the hyphae 5-13 um wide. *Ectal excipulum* thin, of textura globulosa and textura angularis, the exterior cells giving rise to a few minute pustules.

HABITAT-DISTRIBUTION: On conifer duff, Pacific Northwest, fall, common.

DESCRIPTIONS: refs. 14, 31.

REMARKS: Edibility unknown. The very similar *Otidea alutacea* which has hooked paraphyses and a truncate apothecium is distinguished by its grayish brown color and lack of foot like base.

SCARLET ELF CUP
Scutellinia umbrarum (Fr.) Lambotte

FRUITING BODIES: Scattered to gregarious; shallow cupulate, margin hairy, up to 2 cm wide; *hymenium* orange to reddish orange; *exterior* hairy from long brown deep rooting hairs; short superficial hairs scarce or absent.

MICROCHARACTERS: *Spores* 20-24 x 14-16 um, elliptical, rough from scattered warts, hyaline with 1-2 large oil drops at maturity. *Asci* 240-300 x 15-20 um, J-. *Paraphyses* clavate, sometimes branched, up to 9 um at apex. *Medullary excipulum* of textura intricata. Marginal rooting hairs 100-200 um long.

HABITAT-DISTRIBUTION: Common on the ground, rarely on

wood, worldwide, spring to fall.

DESCRIPTIONS: refs. 5, 31, 37.

REMARKS: Edibility untested. The habitat on soil rather than wood, scattered warts on the spores, and the short hairs separate this species from *S. scutellata. Cheilymenia crucipila* of similar habit differs in having fewer rather pale hairs and spores without oil drops. *Scutellinia erinaceus* is distinguished from the above because of the smooth spores.

ELF CUP
Tarzetta cupularis (L. ex Fr.) Lambotte

FRUITING BODIES: Gregarious to crowded; subsessile, cupulate, margin crenate, up to 2 cm wide; *hymenium* grayish tan, smooth; *exterior* concolorous, minutely pustulate.

MICROCHARACTERS: *Spores* 20-23 x 10-12 um, elliptical smooth, hyaline with two large oil drops. *Asci* 260-350 x 14-16 um, J-. *Paraphyses* slightly clavate, occasionally branched and septate and sharply bent, up to 5 um broad at apex. *Medullary excipulum* of densely interwoven textura intricata. *Ectal excipulum* textura angularis to interior and textura globulosa toward exterior, the outer cells heaped to form the pustules.

HABITAT-DISTRIBUTION: Common on soil and among moss in coniferous forests throughout the Pacific Northwest, spring and summer.

DESCRIPTIONS: refs. 6, 8, 9, 31.

REMARKS: Edibility untested. It resembles *Geopyxis* which does not have oil drops in the spores.

GRAY HAIRY ELF CUP
Trichophaea boudieri Grel.

FRUITING BODIES: Scattered to gregarious; sessile, cupulate, then flat, up to 5 mm wide; *hymenium* pale gray; smooth; *exterior* dark brown, with pale brown stiff bristly hairs at margin only; pustulate to short soft-hairy elsewhere.

MICROCHARACTERS: *Spores* 23-25 x 12-13 um, broadly fusoid, smooth, with many oil drops. *Asci* 250-300 x 14-17 um, tapered abruptly at base. *Paraphyses* slightly clavate, the tips up to 4 um wide. *Hypothecium* a dense layer of textura angularis. *Medullary excipulum* of textura angularis the cells of which are larger than those in the hypothecium. *Ectal excipulum* of textura globulosa, the cells giving rise to the excipular dark brown hairs.

HABITAT-DISTRIBUTION: On the ground, duff and litter, spring. Northern Idaho and Cedar-Hemlock and Douglas Fir zones, not common.

DESCRIPTIONS: refs. 11, 15, 31.

REMARKS: Edibility untested. It is too small to be of interest to the pothunter. *Humaria hemispherica* is another species with a whitish hymenium and brown pointed hairs. It is distinguished by the broadly elliptical spores with two conspicuous oil drops, and late summer and fall fruiting. *Trichophaea abundans* occurs on burned soil and has marginal hairs that arise from the ends of chains of excipular cells.

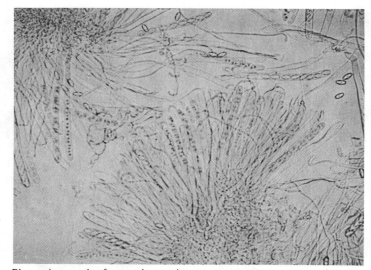

Photomicrograph of operculate asci, ascospores, and paraphyses of *Otidea smithii*. Magnification X200.

TRUFFLES
The Tuberales

These are the famous truffle fungi, which unfortunately are not as well represented in Idaho as they are on the coast of Oregon and California. Furthermore, their habit of growing under ground has led to their neglect so that we know very little about them. Their fruit bodies are typically globose, lobed or wrinkled and are easily confused with the false puffballs or Hymenogastrales which incidentally are exceedingly common in Idaho. Chances are if you encounter a truffle-like fungus it is not a truffle but one of the false puffballs. This can be determined best by a microscopic study which will reveal the presence of asci in the case of a truffle or basidia if one has a false puffball.

Diagnostic Description

Ascocarps hypogean, contents fleshy, waxy, or powdery dry; asci without pore, spores not forcibly discharged; ascospores hyaline or pigmented, smooth, rough, often spherical.

List of Species

1. *Elaphomyces granulatus* Fr.
2. *Hydnotrya cerebriformis* Harkness

Composite photograph of hypogeous fungi: Truffles are easily confused with False Puffballs. The Deer Truffle in center is flanked by False Puffballs *(Gautieria)* on left and *Rhizopogon* on right.

DEER TRUFFLE
Elaphomyces granulatus Fr.

FRUITING BODIES: Scattered to gregarious; globose, 2 to 4 cm wide; dingy ochraceous to yellowish brown, embedded in a crust of soil impregnated with yellowish mycelium, outer surface covered with pyramidal warts, outer wall tough, white in section, central mass purplish black from spores, moist when wet, powdery when dry, at first divided into compartments by thin bands of sterile tissue.

MICROCHARACTERS: *Spores* 24-32 um, globose, blackish-brown, with low irregular warts, wall up to 4 um thick. *Asci* 35-45 um, globose to pyriform, thin-walled, evanescent, 6 spored.
HABITAT-DISTRIBUTION: Widespread, subterranean, frequently embedded in hard pan 2-3 inches under humus, under conifers particularly pine, throughout the growing season; common in the McCall area where it can be collected by the bushel.
DESCRIPTION: ref. 6.
REMARKS: Edible, and known from the sixteenth century as an aphrodisiac. The flavor is said to be inferior to that of the true truffle *Tuber*. Occasionally parasitized specimens are encountered with club-shaped stromata of *Cordyceps capitata* arising from the fruiting bodies.

BRAIN TRUFFLE
Hydnotrya cerebriformis Harkness

FRUITING BODIES: Single to scattered; up to 3 cm wide, fleshy, brown, coarsely lobed, with a fuzzy surface; contents convoluted, pinkish brown of branched canals, with a whitish hymenium, the entire mass appearing brain-like.

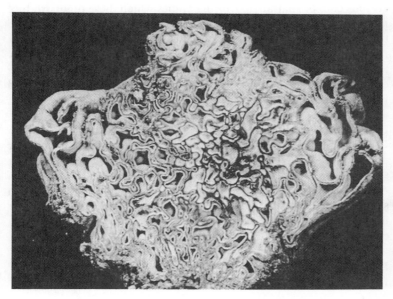

MICROCHARACTERS: *Spores* 25-30 um in diameter delicately roughened. *Asci* cylindric, 8 spored.
HABITAT-DISTRIBUTION: In conifer duff, buried or partially erumpent late spring or summer, rare.
DESCRIPTION: N. Amer. Flora 1954. Series II 1:10.
REMARKS: Edibility untested. A rare fungus collected only twice in northern Idaho.

EARTH TONGUES
The Helotiales - Geoglossaceae

Since the edibility of the earth tongues is essentially untested, this group will be of marginal interest to the mushroom collector. The curious club or spathulate fruiting bodies which these fungi produce might be confused with small simple coral fungi. All of the earth tongues produce asci which have a pore or tear at the apex (inoperculate) and with the exception of *Sprageola* the spores are long - filamentous rather than elliptical or globose as in the operculate fungi. In one group the ascocarps are capitate or knobed rather than clavate or spatulate.

Diagnostic Description
Fruiting bodies relatively large, club-shaped, spatulate, capitate or pileate with the hymenium covering the upper surface, the lower surface sterile, fleshy, fibrous, leathery, or gelatinous, never brittle. Ascospores ejected through an apical pore in the ascus, varying in shape from ellipsoid to fusiform or long, needle-like, hyaline or brown. Species occurs on rotting wood, soil, duff, or decaying conifer needles.

LIST OF SPECIES
1. *Cudonia circinans* Fr.
2. *Cudonia grisea* Mains
3. *Cudonia monticola* Mains
4. *Geoglossum fallax* Durand
5. *Geoglossum glabrum* Fr.
6. *Geoglossum nigritum* (Fr.) Cooke
7. *Leotia lubrica* Fr.
8. *Mitrula abietis* Fr.
9. *Mitrula gracilis* Karsten
10. *Spathularia flavida* Fr.
11. *Spathularia velutipes* Cooke & Farl.
12. *Sprageola irregularis* (Pk.) Nannf.
13. *Trichoglossum hirsutum* (Fr.) Boudier
14. *Trichoglossum velutipes* (Peck) Durand
15. *Vibrissea truncorum* Fr.

KEY

1a Fruiting bodies black, or very dark brownish black, ascospores brown at maturity 2

1b Fruiting bodies, tan, pale to pinkish brown, gray, yellow, not black ascospores hyaline at maturity 5

2a Ascocarp delicately velvety from dark brown setae on stipe and in hymenium; asci 8 spored, apothecia clavate to capitate, compressed above, up to 2 cm long and 2-5 mm wide; ascospores 80-170 x 5-7 um, fusoid-clavate, up to 15 septate; on soil, rotting wood or moss . . . . .
. . . . *Trichoglossum hirsutum*

2b Ascocarps without setae, not velvety 3

3a Ascospore color variable, mixed hyaline and brown; clavate or clavate cylindric 66-90 x 5-6 um, 0-13 septate, hyaline; ascocarps brown, 1-7 cm long and 2-10 mm wide; stipe squamulose; paraphyses not or only somewhat agglutinated by amorphous matter . . . .
. . . . *Geoglossum fallax*

3b Ascospores brown 4

4a Paraphyses with upper cells enlarged: globoid, ellipsoid or obovoid; ascospores 55-78 x 6-8 um, clavate, 7-septate, straight or slightly curved, dark brown; stipe smooth; apothecia black, clavate, up to 10 cm long and 3-8 mm wide; on soil . . . . *Geoglossum glabrum*

4b Paraphyses with upper cells not or only slightly enlarged, strongly curved; ascospores 30-90 x 4.5-6.5 um, regularly 7-septate, straight or somewhat curved, clavate, dark brown; stipe glabrous or minutely pubescent or squamulose; apothecia dark brown to black, clavate, up to 7 cm long and 1-5 mm wide; on moss, soil or soggy stream banks . .
. . . . *Geoglossum nigritum*

5a Ascocarp resembling a deformed, contorted, clavarioid fungus, paraphyses absent, ascospores ovoid to globose, 6-8 x 3-5 um, smooth, 1 celled; ascocarp lemon to cadmium yellow, 1-7 cm long by 2-15 mm wide, stipe white 2-8 mm thick; on ground among moss . . . .
. . . . *Sprageola irregularis*

5b Ascocarps spatulate, capitate, helvelloid, etc., not clavarioid; paraphyses present 6

6a Apothecia compressed, fan-shaped, spatulate, 1-8 cm long up to 3 cm wide, sometimes lobed or contorted, yellow to cinnamon buff; ascospores 45-56 x 2-2.6 um, acicular, rounded above, acuminate below, 0-several septate, wall with gelatinous layer, conidia sometimes filling ascus; paraphyses strongly curved; on ground under conifers, at times in rings and arcs . . . . *Spathularia flavida*

6b Apothecia, stipitate-pileate, capitate, not compressed laterally 7

7a Ascocarps viscid-gelatinous, ochraceous yellow, at times with a greenish tinge, 3-6 cm high, pileate portion furrowed, wrinkled or nodulose; ascospores 18-28 x 5-6 um, cylindric oblong to fusiform, straight or curved, 5-7 septate; stipe viscid, minutely squamulose; on soil or rotting wood . . . . *Leotia lubrica*

7b Ascocarps fleshy, toughish, not gelatinous 8

8a Ascospores over 85 um long, apothecia yellow to orange, knob-like with a brownish stalk, growing on sticks submerged in cool mountain streams; ascospores multiseptate .... *Vibrissea truncorum*

8b Ascospores shorter, not over 70 um long, not growing in streams on submerged wood, ascocarp, brown, flesh or avellaneous colored ...9

9a Apothecia pileate, with a fertile upper portion and a sterile lower surface 10

9b Apothecia capitate, with the enlarged apex completely fertile 12

10a Ascocarps drab or dark gray; ascospores short, up to 24 um long; pileus 0.5-1.5 cm wide, convex, smooth; stipe fuscous, terete, smooth 3-8 mm thick; ascospores 18-24 x 1.5-2 um, 1 celled, rarely once septate; on rotting coniferous wood........................ *Cudonia grisea*

10b Pileus cream to dark brown or pinkish buff 11

11a Pileus cream, stipe furfuraceous, striate to ridged, drab to dark brown; ascocarps up to 7 cm long and 2 cm wide, convex, smooth, wrinkled or sometimes convoluted; ascospores 32-40 x 2 um, acicular, aseptate or sometimes several septate, wall, thin, gelatinous; on soil, less often on wood .... *Cudonia circinans*

11b Pileus pinkish buff to grayish brown; stipe glabrous, avellaneous, somewhat compressed; ascospores 18-24 x 2 um, 1 celled, rarely septate, acicular to narrowly clavate, on the ground and on rotting wood .... *Cudonia monticola*

12a Heads light brown to pinkish buff; stipe darker than head, dark brown; ascospores 10-14 x 2-2.5 um; on conifer needles ..... *Mitrula abietis*

12b Heads ochraceous to orange buff, rugose, to cerebriform; stipe lighter than head, creamy white; ascospores 9-12 x 1.5-2 um; on ground in mosses....................................... *Mitrula gracilis*

Mycophagy: Morels and other Discomycetes are best preserved by canning (usually in brine) because both flavor and texture are degraded upon freezing or drying in the usual way.

MOUNTAIN LOVING CUDONIA
Cudonia monticola Mains

FRUITING BODIES: Cespitose to densely gregarious; pileate-stipitate, fleshy, leathery, up to 10 cm long and 3 cm wide; *pileus* pinkish cinnamon to grayish brown, laterally compressed, irregularly hemispheric, or helvelloid, margin inrolled; *stipe* pallid grayish brown, glabrous, hollow at maturity, up to 7 mm wide at base, narrowing toward apex.

MICROCHARACTERS: *Spores* 18-24 x 2 um, needle-like to narrowly clavate, rarely one septate; *paraphyses* hyaline, curved to uncinate, filiform.

HABITAT-DISTRIBUTION: On soil and rotting conifer wood in spring and summer, at times in snow banks. Common in Idaho and Pacific Northwest.

DESCRIPTION: ref. 26.

REMARKS: Edibility untested. This species differs from *Cudonia circinans* which it very much resembles, by its smaller ascospores and spring fruiting. It is a common element of the snow bank flora, being found at times fruiting in the spring on logs buried in or under the snow. *Cudonia grisea* also with short ascospores is distinguished by the dark gray to drab pileus and fuscous stipe.

BLACK EARTH TONGUE
Geoglossum nigritum (Fr.) Cooke

FRUITING BODIES: Scattered; black to dark brown, clavate, fertile portion occupying upper one half to one third, somewhat compressed, at times protruding asci give the surface a velvety appearance, lacking setose dark hairs; *stipe* concolorous, terete, glabrous to furfuraceous or squamulose, slender, up to 2 mm thick.

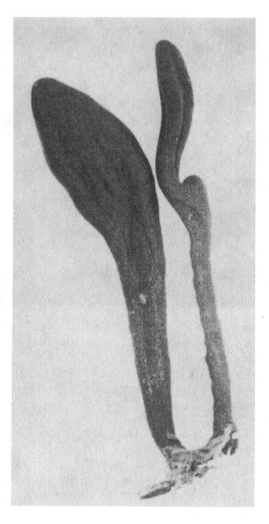

MICROCHARACTERS: *Spores* 30-90 x 4.5-6.5 um, clavate; slightly curved, dark brown, typically 7-septate; *Asci* 125-180 x 16-20 um, clavate, 8 or less spored. *Paraphyses*, hyaline to brown, some enlarged at apex, the terminal cell clavate, narrowly obovoid or cylindric, many curved.

HABITAT-DISTRIBUTION: Growing among moss or on rotting wood under conifers, summer and fall, widespread, but only occasionally encountered.

DESCRIPTION: ref. 24.

REMARKS: Edibility untested. A number of similar dark earth tongues such as *G. fallax* or *G. glabrum* are scarcely distinguishable on a field basis. In using the key provided, microscopic examination is needed to separate them.

SLIPPERY LEOTIA
Leotia lubrica Fr.

FRUITING BODIES: Gregarious to clustered; yellowish green to olivaceous, up to 6 cm high, fertile head convex, furrowed or wrinkled, margin lobed and overhanging, squamulose below; *stipe* ochraceous, dotted with greenish granules, tapering upward, 5-10 mm thick, consisting of three layers, the inner and outer layers gelatinous.

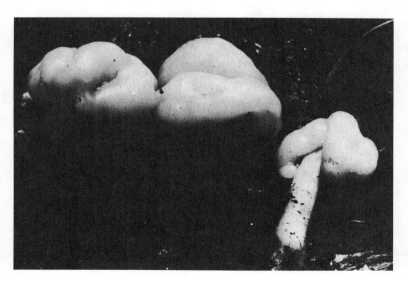

MICROCHARACTERS: *Spores* 16-23 x 4-6 um, subfusoid with rounded ends, hyaline with many oil drops, becoming 5-7 septate. *Asci* 115-150 x 7-10 um, clavate. *Paraphyses* branched below, enlarged to tips which are often agglutinated with amorphous material.

HABITAT-DISTRIBUTION: On soil or rotting wood, summer, fall, widespread but occasionally encountered.

DESCRIPTIONS: refs. 6, 26.

REMARKS: Edibility untested. The gelatinous tissue and viscid surface readily distinguish it from any *Cudonia* with which it may be confused.

FIR-NEEDLE MITRULA
Mitrula abietis Fr.

FRUITING BODIES: Scattered to gregarious; capitate, up to 2 cm long, usually smaller, heads sharply delimited from stipe, cylindric to hemispheric light brown to pinkish buff; *stipe* light to dark brown, somewhat pruinose above, and with brownish tomentum at base.

MICROCHARACTERS: *Spores* 10-14 x 2-2.5 um, fusoid, one-celled. *Asci* 50-70 x 4-6 um, clavate. *Paraphyses* clavate, hyaline, branched below.

HABITAT-DISTRIBUTION: On decaying conifer needles, fall, Rocky Mountains and Pacific Northwest.
DESCRIPTIONS: refs. 6, 25.
REMARKS: Edibility untested. This is a very small inconspicuous species which blends in with the fallen needles on which it grows. *Mitrula gracilis* grows among mosses and has an ochraceous to orange buff head.

YELLOW EARTH TONGUE
Spathularia flavida Fr.

FRUITING BODIES: Scattered to gregarious at times in rings and arcs; spatulate, up to 8 cm high, light to strong yellow, the flattened fertile area at times paler; fertile area often irregularly wrinkled and sometimes notched at apex, up to 2 cm wide; *stipe* hollow, glabrous, with white to yellowish mycelium at base.

MICROCHARACTERS: *Spores* 30-95 x 1.5-2.5 um, non-or several septate, acicular, wall with gelatinous layer. *Asci* 85-125 x 8-12 um, clavate. *Paraphyses* filiform, hyaline, some circinate above.
HABITAT-DISTRIBUTION: Gregarious on duff under conifers, summer and fall, common in Pacific Northwest.

DESCRIPTIONS: refs. 6, 25.
REMARKS: Edibility untested; the small size would discourage table use. *Spathularia velutipes* has a brown velvety stipe and orange mycelium.

VELVETY-BLACK EARTH TONGUE
Trichoglossum hirsutum (Fr.) Boudier

FRUITING BODIES: Scattered to gregarious; clavate, at times forked or branched, brownish black up to 8 cm long; *fertile portion* velvety from setae, compressed up to 2 cm long and up to 5 mm wide; *stipe* concolorous, velvety, up to 3 mm wide.

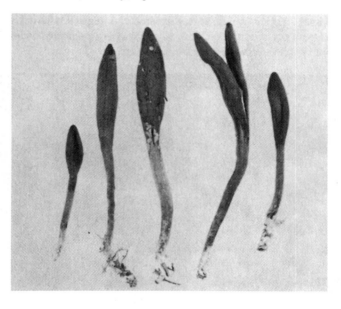

MICROCHARACTERS: *Spores* 80-170 x 5-7 um, fusoid-clavate, often 15-septate, brown. *Asci* 180-275 x 18-25 um, clavate, pore bluing in iodine, 8-spored. *Paraphyses* brown, filiform, somewhat enlarged and often curved and brown near apex; *setae* dark brown, opaque, pointed, projecting up to one-third their length.
HABITAT-DISTRIBUTION: On moss, soil or rotting wood under conifers, summer and fall, widespread, but only occasionally encountered.

DESCRIPTIONS: refs. 6, 24.
REMARKS: Edibility untested. *T. velutipes* is very similar but has 4 spored asci.

AQUATIC EARTH TONGUE
Vibrissea truncorum Fr.

FRUITING BODIES: Scattered, gregarious to cespitose; with knoblike fertile head, fleshy to gelatinous, yellowish orange, drying reddish orange, up to 1.5 cm high and 2.5 mm wide, sterile underneath; *stipe* with short downy blackish hairs.

MICROCHARACTERS: *Spores* 125-250 x 1 um, thread-like, multiseptate. *Asci* variable about 300 x 5-6 um, cylindrical, 8 spored. *Paraphyses* filiform below, enlarged at apex, simple or forked near tip.
HABITAT-DISTRIBUTION: On sticks and wood submerged in cold running water, summer, widespread but seldom encountered.
DESCRIPTION: ref. 26.
REMARKS: Edibility untested. Often encountered in the mountains in cold running water. The ascospores, when mature, project and vibrate in the running water above the fertile area giving the fruiting body a white silky appearance.

Mushroom habitat: The edges of forests often yield Walnuts early in spring and Morels later. Shown is Strychnine Creek area, Latah County, Idaho.

REFERENCES

1. BERTHET, P. and L. RIOUSSET. 1965. Un *Urnula* Nouveau des cedrais provencales: *Urnula pouchetii* Nov. sp. (Discomycetes opercules). Bull. Mens. Soc. Linn. Lyon 34: 263-261.

2. BRUMMELEN, J. VAN. 1967. A world-monograph of the genera *Ascobolus* and *Saccobolus*. Persoonia. Suppl. I: 1-260.

3. BURDSALL, H.H., Jr. 1968. A revision of the genus *Hydnocystis* (Tuberales) and of the hypogeous species of *Geopora* (Pezizales). Mycologia 60: 496-525.

4. DENISON, W.C. 1964. The genus *Cheilymenia* in North America. Mycologia 56: 718-737.

5. DENISON, W.C. 1959. Some species of the genus *Scutellinia*. Mycologia 51: 605-635.

6. DENNIS, R.W.G. 1968. British Ascomycetes. Cramer, Lehre. 445 p.

7. DISSING, H. 1966. The genus *Helvella* in Europe, with special emphasis on the species found in Norden. Dansk Bot. Ark. 25(1): 1-172.

8. DURAND, E.J. 1900. The Classification of the fleshy Pezizinae with reference to the structural characters illustrating the basis of their division into families. Bull. Torr. Bot. Club 27: 463-495.

9. ECKBLAD, F.E. 1968. The genera of the operculate Discomycetes. A re-evaluation of their taxonomy, phylogeny and nomenclature. Nyatt Mag. Bot. 15: 1-191.

10. ELLIOTT, M.E., and M. KAUFERT. 1974. *Peziza badia* and *Peziza badio-confusa*. Canad. J. Bot. 52: 467-472.

11. GRELET, L.J. 1917. Un discomycete noveau, le *Trichophaea boudieri* sp. nov. Bull. Soc. Mycol. France 33: 94-96.

12. KANOUSE, B.B., and A.H. SMITH. 1940. Two new genera of Discomycetes from the Olympic National Forest. Mycologia 32: 756-759.

13. KANOUSE, B.B. 1948. The genus *Plectania* and its segregates in North America. Mycologia 40: 482-497.

14. KANOUSE, B.B. 1949. Studies in the genus *Otidea*. Mycologia 41: 660-677.

15. KANOUSE, B.B. 1958. Some species of the genus *Trichophaea*. Mycologia 50: 121-140.

16. KEMPTON, P.E., and V.L. WELLS. 1970. Studies on the fleshy fungi of Alaska. IV. A preliminary account of the genus *Helvella*. Mycologia 62: 940-959.

17. KEMPTON, P.E., and V.L. WELLS, 1973. Studies on the fleshy fungi of Alaska. VI. Notes on *Gyromitra*. Mycologia 65: 396-400.

18. KIMBROUGH, J. 1969. North American species of *Thecotheus* (Pezizales, Pezizaceae). Mycologia 61: 99-114.

19. KIMBROUGH, J., E.R. LUCK-ALLEN, and R.F. CAIN. 1969. *Iodophanus*, the Pezizeae segregate of *Ascophanus* (Pezizales). Amer. J. Bot. 56: 1187-1202.

20. KIMBROUGH, J., E.R. LUCK-ALLEN, and R.F. CAIN. 1972. North American species of *Coprotus* Thelebolacceae. (Pezizales). Canad. J. Bot. 50: 957-971.

21. KORF, R.P. 1957. Two bulgarioid genera: *Galiella* and *Plectania*. Mycologia 49: 107-111.

22. LARSEN, H.J., Jr. and W.C. DENISON. 1978. A checklist of the operculate cupfungi (Pezizales) of North America West of the Great Plains. Mycotaxon 7: 68-90.

23. LINCOFF, G. and D.H. MITCHEL. 1977. Toxic and hallucinogenic mushroom poisoning. Van Nostrand Reinhold Co., New York. 267 p.

24. MAINS, E.B. 1954. North American species of *Geoglossum* and *Trichoglossum* Mycologia 46: 586-631.

25. MAINS, E.B. 1955. North American hyaline-spored species of Geoglossaceae. Mycologia 47: 846-877.

26. MAINS, E.B. 1956. North American species of Geoglossaceae. Tribe Cudonieae. Mycologia 48: 694-710.

27. McKNIGHT, K.H. 1969. A note on *Discina*. Mycologia 61: 614-630.

28. MILLER, O.K., Jr. 1967. Notes on western Fungi. I. Mycologia 59: 504-512.

29. MILLER, O.K., Jr. 1972. Mushrooms of North America E.P. Dutton & Co. Inc. N.Y. 360 p.

30. MOORE, E.J., and R.P. KORF. 1963. The genus *Pyronema*. Bull. Torrey Bot. Club 90: 33-43.

31. PADEN, J.W. 1967. A taxonomic study of the Pezizales of northern and central Idaho. Ph.D. Dissertation U. of Idaho. 247 p.

32. PADEN, J.W., J.R. SUTHERLAND, and T.A.D. WOODS. 1978. *Caloscypha fulgens* (Ascomycetidae, Pezizales): the perfect state of the conifer seed pathogen *Geniculodenron pyriforme* (Deuteromycotina, Hyphomycetes). Canad. J. Bot. 56: 2375-2379.

33. PADEN, J.W. and E.E. TYLUTKI. 1968. Idaho Discomycetes. I. Mycologia 60: 1160-1168.

34. PADEN, J.W. and E.E. TYLUTKI. 1969. Idaho Discomycetes. II. Mycologia 61: 683-693.

35. PADEN, J.W. 1972. A new combination in *Neournula*. Mycologia 64: 457.

36. PFISTER, D.H. 1973. The psilopezioid fungi. IV. The genus *Pachyella* (Pezizales). Canad. J. Bot. 51: 2009-2023.

37. SEAVER, F.J. 1942. The North American cup-fungi (operculates). Suppl. ed. Seaver, New York. 377 p.; *reprinted* 1961. Hafner, New York.

38. SEAVER, F.J. 1951. The North American cup-fungi (inoperculates) Published by the author N.Y. 428 p.

39. SMITH, A.H. 1975. A field guide to western mushrooms. Univ. of Michigan Press. Ann Arbor. 280 p.

40. SMITH, A.H. 1948. Mushrooms in their natural habitats. Sawyers Inc. Portland 626 p.

41. SNYDER, L.C. 1936. New and unusual discomycetes of western Washington. Mycologia 28: 483-488.
42. SYNDER, L.C. 1938. The operculate discomycetes of western Washington. Univ. of Washington Pub. in Biol. 8: 1-64.
43. WEBER, N.S. 1972. The genus *Helvella* in Michigan. Mich. Botanist 11: 147-201.
44. WELLS, V., and P. KEMPTON. 1967. Studies of the fleshy fungi of Alaska. I. Lloydia 30: 258-268.

Mushroom collecting: Mushrooming is family fun. Children love it, are eager to learn, and are a real help in locating mushrooms that might otherwise be passed by. Shown above is the author's family of helpers.

PLATE I

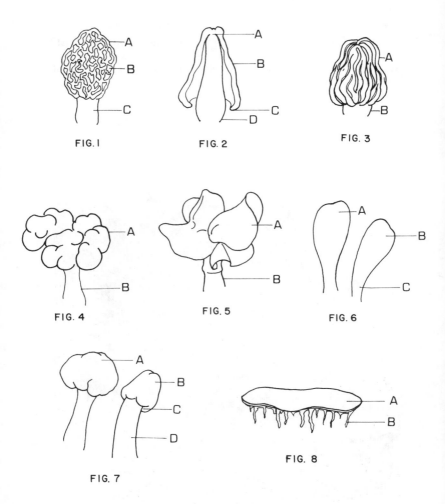

FIG. I

FIG. 2

FIG. 3

FIG. 4

FIG. 5

FIG. 6

FIG. 7

FIG. 8

Fig. 1   *Morchella esculenta* A. Pileus B. Hymenium C. Stipe
Fig. 2   *Ptychoverpa bohemica* A. cross section of pileus showing attachment at apex only B. Hymenium C. Margin free D. stipe
Fig. 3   *Ptychoverpa bohemica* A. pileus with wrinkled surface B. Stipe
Fig. 4.   *Gyromitra esculenta* A. Brain-like or gyrose pileus B. Stipe
Fig. 5   *Gyromitra infula* A. Saddle-shaped pileus B. Stipe
Fig. 6   *Spathularia flavida* A. Flattened apex B. Hymenium C. Stipe
Fig. 7   *Cudonia monticola* A. Capitate pileus B. Hymenium C. Margin attached D. Stipe
Fig. 8   *Rhizina undulata* A. Hymenium B. Rhizoid-like strands

GLOSSARY

ACYANOPHILIC — not staining blue with the dye cotton blue.

AMORPHOUS — without any definite shape.

AMYLOID — staining blue, bluish black or grayish blue with Melzer's Reagent.

ANASTOMOSING — joined or branching in the manner of a network.

APICAL. of or pertaining to the apex or tip.

APOTHECIUM the fruiting body of one of the discomycetes, Fig. 24

ASCOCARP the fruiting body of an ascomycete.

ASCOSPORE a spore produced within an ascus by free cell formation, Fig. 19.

ASCUS the microscopic cylindric, clavate or globose cell in which the ascospores are produced. Fig. 19.

AVELLANEOUS a color term meaning pale grayish brown or hazel.

BASIDIUM the cell in which karyogamy and meiosis occurs and on which the basidiospores (typically four) are produced.

CAMPANULATE — bell-shaped.

CAROTENOID — containing carotene pigments which are colored bright red, orange, or yellow.

CATENULATE — borne in chains.

CEREBRIFORM — convoluted or contorted and folded like a brain, Fig. 4.

CLAVARIOID — shaped like a coral fungus.

CLAVATE — club-shaped, (of paraphyses) wider toward the apex, Fig. 6.

CRENATE — scalloped.

CYANOPHILIC — staining blue with the dye cotton blue.

DISCOID — dish shaped.

ECTAL EXCIPULUM — the outer sterile tissue of the apothecium external to the medullary excipulum, Fig. 24.

EGUTTULATE — without oil drops, Fig. 22.

ELLIPTIC - ELLIPSOID — shaped like an ellipse, Fig. 20.

EPIGEOUS - EPIGEAN — growing on or above the ground.

EXCIPULUM (-AR) — the sterile tissue of an apothecium below the level of the hymenium and hypothecium, if present, Fig. 24.

EXTERIOR — The external non-spore bearing portion of the fruitbody in a cup fungus. In a repand, saddle-shaped or *Helvella* type fruiting body the exterior is the lower or innermost least exposed surface.

PLATE II

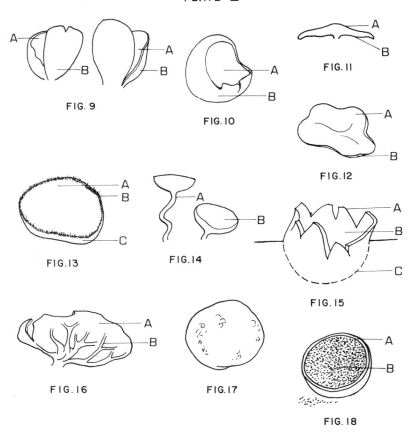

FIG. 9

FIG. 10

FIG. 11

FIG. 12

FIG. 13

FIG. 14

FIG. 15

FIG. 16

FIG. 17

FIG. 18

Fig. 9   *Otidea alutacea:* Otidioid fruiting body, ascocarp cleft nearly to base and apex truncate. A. Hymenium B. Exterior

Fig. 10   *Calosypha fulgens:* Deep cupulate fruiting body, margin split. A. Hymenium B. Exterior.

Fig. 11   *Discina perlata:* Cross section of fruiting body to show repand condition. A. Hymenium B. Exterior

Fig. 12   *Peziza sylvestris:* Shallow cupulate or discoid fruiting body with wrinkled hymenium. A. Hymenium B. Exterior

Fig. 13   *Scutellinia scutellata:* Hairy apothecium. A. Hymenium B. Hairy or setose margin C. Exterior

Fig. 14   *Plectania nannfeldtii:* Long stipitate apothecium. A. Stipe B. Hymenium

Fig. 15   *Sarcosphaera crassa:* Emersed, deep cupulate apothecium with stellate margin. A. Margin B. Hymenium C. Hypogeous base.

Fig. 16   *Helvella acetabulum:* Apothecium with ribs. A. Exterior B. Ribs

Fig. 17   *Elaphomyces granulatus:* Globose truffle-type ascocarp

Fig. 18   *Elaphomyces granulatus:* Ascocarp sectioned to show contents. A. Rind or Cortex B. Powdery spore mass when mature.

FILIFORM — long filamentous, or thread-like, Fig. 21.

FRUITING BODY — an organized mass of fungal tissue in or on which spores are produced as a result of sexual reproduction. The structure is commonly referred to as a mushroom if large and conspicuous. In the cup fungi (Discomycetes) the fruiting body is called an apothecium.

FURFURACEOUS — covered with small flaky granules varying in size somewhere between sugar and flour, scurfy.

FUSOID (FUSIFORM) — spindle-like, tapered at both ends.

GLABRESCENT — hairy at first then smooth or glabrous.

GLABROUS — smooth, without hairs or other types of vestiture.

GLOBOSE — round, spherical, or shaped like a globe.

GUTTULATE — containing oil drops, Fig. 23.

HABITAT — the place where a fungus is found.

HAIR — a generalized term for a single or multiple-hyphal appendage which is attached at one end and generally tapered to a narrow-free end.

HELVELLOID — shaped like the fruiting body of one of the *Helvella* species which is typically stipitate with a saddle or miter-shaped pileus.

HYALINE — clear, glassy or transparent as seen under the microscope, not pigmented.

HYMENIUM — in Ascomycetes, the fertile area of a fruiting body where asci and ascospores are produced, often differing in color, texture or topography from the remainder.

HYPOGEOUS — subterranean, fruiting below the surface of the ground.

HYPOTHECIUM — the tissue immediately below the layer of asci in an apothecium and subtended by the medullary excipulum, subhymenium, Fig. 24.

INOPERCULATE — the type of ascus which lacks an operculum but has a pore or irregular tear through which the spores are discharged, Fig. 21.

LIGNICOLOUS — growing on wood.

MARGIN — the area of an apothecium where the hymenium ends and the sterile tissue begins — the edge of the fertile area of an apothecium.

MEDULLARY EXCIPULUM — the region of sterile tissue of the apothecium between the hymenium (or hypothecium, if differentiated) and the ectal excipulum, Fig. 24.

MELZER'S REAGENT — An iodine containing reagent used for staining starch or starch-like components of fungal hyphae and spores. Contains: 1.5 g iodine, 5 g potassium iodide, 100 g chloral hydrate, and 100 g water.

PLATE III

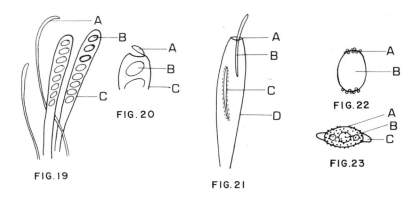

FIG.19   FIG.20   FIG.21   FIG.22   FIG.23

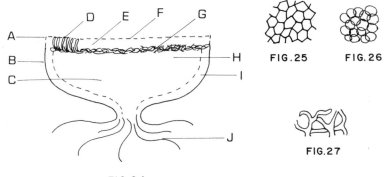

FIG.24   FIG.25   FIG.26   FIG.27

Fig. 19   *Otidea smithii:* Asci and paraphyses A. Paraphysis B. Elliptical asco-spore C. Ascus
Fig. 20   Same showing operculum: A. Operculum B. Ascospore C. Ascus
Fig. 21   *Spathularia flavida:* Inoperculate ascus. A. Pore torn in apex as spores are discharged. B. Needle-like ascospores C. Gelatinous sheath surrounding ascospore D. Ascus
Fig. 22   *Morchella esculenta:* Ascospore A. Polar exterior granules B. Eguttulate spore contents.
Fig. 23   *Discina perlata:* Apiculate verrucose ascospore. A. Spore ornamentation B. Oil Drop or guttule C. Apiculus
Fig. 24   X-sect. diagram of typical apothecium. A. Margin B. Exterior C. Flesh or context D. Ascus E. Ascus layer or hymenium F. Interior or hymenium G. Subhymenium H. Medullary excipulum I. Ectal excipulum J. Mycelium
Fig. 25-27 Types of Excipulum
Fig. 25   Textura angularis
Fig. 26   Textura globulosa
Fig. 27   Textura intricata

126

MICROCHARACTERS — those features of a mushroom which are determined using the magnifications obtained with a light microscope.

MITRATE (MITRULATE) — shaped like a bishops miter.

NON-SEPTATE — term used to designate mycelium which does not have cross walls.

OBOVOID — egg-shaped with the widest portion toward the apex or upper side, the inverse of ovoid.

OCHRACEOUS — a color term meaning yellowish of about the shade of corn (maize).

OPERCULUM (OPERCULATE) — the hinged trapdoor-like apex on the ascus of certain discomycetes. The operculum is pushed open upon spore discharge, Fig. 20.

OTIDIOID — a fruiting body shaped like a species in the genus *OTIDEA* i.e. split on one side nearly to base, spoon-shaped, cleft, etc., Fig. 9.

PARAPHYSES — filamentous sterile elements scattered among the asci in a hymenium, Fig. 19.

PARENCHYMATOUS — fungal tissue in which the element are globose or soap bubble like, and not filamentous, Fig. 26.

PILEUS — that portion of a fruitbody which is abruptly defined from the stipe by shape, color or texture and which is typically fertile on one side only, the cap. Fig. 1.

PROSENCHYMATE (-OUS) — a fungal tissue in which the filamentous nature of the elements is clearly evident.

PULVINATE — humped or cushion-shaped.

PUSTULES — small raised granular packettes of cells on the external surface of an apothecium, giving the surface a blistery or rough texture.

REPAND — the shape of a discoid apothecium having a downward curved, wavy margin, Fig. 11.

RHIZOIDS — root-like strands of hyphae at the base or lower side of the fruiting body of *Rhizina*, Fig. 8.

RIBS — sharp creases or folds that extend from the surface of the fruit body typically on the stipe, Fig. 16.

RIND — the tough, usually thick outer layer of a fruiting body as in *Elaphomyces*, the cortex, Fig. 18.

RUFOUS — dull red.

RUGOSE — wrinkled.

SAPROBES — organisms that grow on dead organic matter.

SCLEROTIUM — a hard compact somatic mass of fungal tissue which is resistant to adverse environmental condition.

SEPTATE — term used to designate mycelium which has cross walls.

SESSILE — without a stalk or stipe, Fig. 10.

SETAE — stiff long bristle-like hairs, which are usually dark colored, Fig. 13.

SPORE — a fungus propagule-in the Discomycetes, the ascospores, Fig. 20.

SQUAMULOSE — adorned with small scales.

STELLATE — star-shaped, splitting in a radial fashion to yield star-like points, Fig. 15.

STIPE — the stalk portion of a fruiting body on which the pileus (cap) is born, Fig. 1.

STRIATE — Streaked, usually with thin, straight, nearly parallel lines, grooves or ridges.

SUBALLANTOID — approaching the shape called allantoid, or sausage shaped, curved with rounded ends.

SUBICULUM — a wooly, felty mat of mycelium on which fruiting bodies are formed.

SUBOPERCULATE — an operculate ascus with an interrupted ring at apex.

TETRANUCLEATE — having 4 nuclei.

TEXTURA ANGULARIS — a term used for the tissue of the excipulum which lacks interhyphal spaces and consists of non-filamentous polyhedral elements, Fig. 25.

TEXTURA GLOBULOSA — a term used for the tissue of the excipulum which contains interhyphal spaces and consists of elements which are globose or nearly so, Fig. 26.

TEXTURA INTRICATA — a term used for the tissue of the excipulum in which the elements are filamentous and intricately interwoven, not united along the walls, and are separated by interhyphal spaces, Fig. 27.

TOMENTUM — a soft matted hairy vestiture.

TURBINATE — top-shaped, with the apical region broader than the basal region.

UMBRICAULIFORM — umbrella-shaped.

URCEOLATE — urn-shaped, shaped like a jug or pitcher.

VERRUCOSE — roughened with warts or irregular knobs, Fig. 23.

VINACEOUS — a color term meaning wine colored.

VISCID — a surface which is slimy, slippery or tacky.

INDEX TO COMMON NAMES

INDEX TO SCIENTIFIC NAMES
(Boldface indicates photograph)

131

repanda, Peziza 17, 45, 47, **53**, 56
Rhizina undulata 15, 19, 62, **84**
rutilans, Leucoscypha 85, 90
Saccobolus depauperatus 43
Saccobolus versicolor 43
Sarcoscypha coccinea 31, 33
Sarcosyphaceae 20
Sarcosoma latahensis 17, 21, **29**, 30
Sarcosoma mexicana 16, 21, **30**
Sarcosomataceae 19
Sarcosphaera crassa 13, 15, 19, 44
  45, **60**
scutellata, Scutellinia 14, 86, 87,
  88, 102
Scutellinia erinaceus 86, 102
Scutellinia scutellata 14, 86, 87, 88,
  102
Scutellinia umbrarum 14, 86, 87,
  88, **101**
semilibera, Mitrophora 40
semilibra, Morchella 33, 34, 38
silvicola, Wynnella 17, 62, 100
smithii, Otidea 86, 88, **100**
Spathularia flavida 14, 108, 109,
  115
Spathularia velutipes 108, 116, 117
spinulosa, Lamprospora 85, 89
Sprageola 108
Sprageola irregularis 108, 109
stercoraria, Cheilymenia 85, 89
sylvestris, Peziza 17, 45, 47, **54**
Tarzetta cupularis 86, 87, 90, 96,
  97, **102**
Thecotheus cinereus 43, 44

theloboloides, Cheilymenia 85, 89,
  91
trachycarpa, Plicaria 45, 46, 58, **59**
Tricharina gilva 86, 91
Trichophaea abundans 86, 104
Trichophaea boudierii 86, 90
Trichoglossum hirsutum 14, 108,
  109
Trichoglossum velutipes 108
truncorum, Vibrissea 13, 108, 110,
  117
Tuber 106
Tuberales 18, 96, **105**
umbrarum, Scutellinia 14, 86, 87,
  88, **101**
undulata, Rhizina 15, 19, 62, **84**
Urnula craterium 19, 23
Urnula pouchetii 23
varia, Peziza 17, 45, 47, 53, 55
velutipes, Spathularia 108, 116,
  117
velutipes, Trichoglossum 108
versicolor, Saccobolus 43
venosa, Disciotis 16, 33, 34, **35**
Verpa conica 33, 34, **42**
vesciculosa, Peziza 16, 45, 47, **56**
violacea, Peziza 16, 45, 46, 52, **57**
Vibrissea truncorum 13, 108, 110,
  117
villosa, Helvella 62, 64, 80, **81**
violacea, Peziza 16, 45, 46, 52, **57**
vulcanalis, Geopyxis 85, 87, 91, 96,
  97
vulgaris, Pithya 14, 31, **32**, 89
Wynnella silvicola 17, 62, 100